SAUNDERS MATHEMATICS BOOKS

Consulting Editor

BERNARD R. GELBAUM, University of California

INTRODUCTION TO TOPOLOGICAL GROUPS

TAQDIR HUSAIN *McMaster University*
Hamilton, Ontario

W. B. SAUNDERS COMPANY
Philadelphia and London   1966

Introduction to Topological Groups

PREFACE

This book grew out of my lecture notes mimeographed and distributed by the Department of Mathematics, University of Ottawa, during 1963 and 1964. The enthusiastic reception of these notes by the mathematical community encouraged me to rewrite them and correct minor inaccuracies. Thus, they have been revised and enlarged to the extent that the interested reader will find them self-contained, if he is well prepared in each of the following courses: topology, measure theory on locally compact spaces, groups and linear algebra. Indeed, it is also assumed that the reader is familiar with the elementary concepts of set theory, elements of functional analysis, functions of real and complex variables, and the theory of functions of several variables, especially the Jacobian and Riemann integration.

The present book is probably suitable for a graduate course on topological groups, since it deals with a variety of topics of great importance for understanding the subject and since, without dragging the reader into the blue depths of the subject, it leads him to the foreground where a lot of active research is being done.

Much of the material contained in this book overlaps with other treatises, old and new, on this subject, as one can discern from the contents. However, there are two novel points that merit special mention.

First, unlike other books (e.g., Pontrjagin[37] and Weil[47] among old ones, and Hewitt and Ross[19] among new ones), this books begins with a study of semitopological groups rather than topological groups.

While a topological group is an algebraic group endowed with a topology so that the group operations (viz., multiplication and inversion) are continuous in all variables together, a semitopological group is an algebraic group endowed with a topology so that only the multiplication is continuous in each variable separately. It is quite clear that a topological group is a semitopological group, but the converse is not true. The entire Chapter II of this book has been devoted to finding certain conditions under which the converse is true.

Although some results about when a semitopological group is a topological group have been known for a long time, the existing books on topological groups hardly mention such theorems.

The second novel point of this book lies in Chapter V, which contains very recent material concerning the open homomorphism and the closed graph theorems. Here, indeed, I was motivated by my own work on these theorems in topological vector spaces, where these theorems play a very important role. For details of these theorems on locally convex spaces, the reader is referred to my book, "The Open Mapping and Closed Graph Theorems in Topological Vector Spaces."[22]

The book is organized in such a way that Chapters I to V deal with the algebraico-topological aspect of the subject and Chapters VI to IX emphasize its analytical aspect. This organization has two main advantages: First it is the logical development of the subject. Second, one can select material for one's course according to one's own interests—algebraico-topological or analytical—without much difficulty.

To make the terms commonly used in topology and group theory easily accessible to the reader, Chapter I contains relevant definitions and theorems needed in the sequel. In Chapter II, semitopological groups are studied. Chapter III deals with the general theory of topological groups. In Chapter IV, an elementary study is made of locally compact topological groups. In Chapter V, the open homomorphism and closed graph theorems are proved in a very general setting. From these general theorems, all particular cases known so far are derived.

From Chapter VI on, rudiments of analysis on topological groups are discussed. More specifically, Chapter VI deals with the existence and essential uniqueness of the Haar integral on locally compact groups. The existence and uniqueness proofs given here are due to Weil.[47] Cartan's[8] proof is also mentioned. In Chapter VII, finite-dimensional representations of topological groups are discussed. The famous theorem of Peter-Weyl on metrizable compact topological groups is proved. The proof of this theorem, following Pontrjagin,[37] is given via integral equations. The relevant theorems concerning integral equations are proved. The general theory of representations of a compact group by operators on a Hilbert space is also discussed in exercises at the end of Chapter VII. Chapter VIII deals with the concept of the dual group of a locally compact abelian topological group. A few applications of duality theory are also given. For example, the Plancherel theorem is proved. In Chapter IX, the elementary theory of Banach algebras is introduced. The purpose of this chapter is to show how the theory of Banach algebras subsumes the group algebra $L_1(G)$. This is by no means an exhaustive chapter. For further information, the reader may consult Loomis,[29] among other books.

Finally, it is a pleasure to thank Professor B. Banaschewski for reading the manuscript critically. Moreover, my deepest thanks go to my wife, Martha, for typing the earlier versions of the draft as well as for her moral support. My thanks are also due to Miss Judi Feldman and other secretaries of the Department of Mathematics of McMaster University for typing the final draft so carefully, and to the editors—especially consulting editor, Professor B. Gelbaum for his valuable remarks and suggestions—and the staff of the W. B. Saunders Company for editing and publishing this book.

McMaster University   Taqdir Husain
Hamilton, Ontario

CONTENTS

Fundamentals
of
Topology
and
Group Theory

In this chapter we collect the relevant definitions and results from topology and group theory to make this book self-contained and easy to read. The material is, indeed, standard and can be found in Bourbaki,[4] Kelley,[27] and Van der Waerden.[45] We shall assume that the reader is familiar with common terms used in set theory (e.g., see Abian[1]).

1. TOPOLOGICAL SPACES

A set X with a family u of its subsets is called a *topological space* if the following conditions are satisfied: (a) X and $\varnothing$ (null set) are in u; (b) the intersection of any finite number of members of u is in u; (c) the arbitrary union of members of u is in u.

The members of u are called *u-open* sets of X (or simply *open* sets of X, if there is no topology other than u in question). A topological space X with a topology u will be denoted by X_u.

For any given set X, there are always two topologies on X. These are: (i) u consisting of all subsets of X. It is easy to check that (a)–(c) are satisfied. This topology is called the *discrete topology* on X and denoted by d. X_d is called the *discrete space*.

(ii) u consisting of only X and $\emptyset$. This topology i is called *indiscrete*, and X_i is called the *indiscrete space*.

Let u and v be two topologies on a set X. u is said to be *finer* than v or, in symbols, $u \supset v$ if every v-open set is u-open. If $u \supset v$ then v is said to be *coarser* than u, or equivalently in symbols, $v \subset u$. Furthermore, $u \supset v$ and $v \supset u$ if, and only if, $u = v$ (i.e., u is equal to v). Clearly, d is the finest and i the coarsest topology on any set. Any other topology on a set is finer than i and coarser than d.

Let X_u be a topological space and A any subset of X. The largest open set contained in A is called the *interior* A^0 of A. Clearly, a subset A of X is open if, and only if, $A = A^0$.

The complement $X \sim U$ of an open set U in a topological space X is said to be a *u-closed* or simply a *closed* set. Using the well-known De Morgan Laws in set theory, one verifies the following: (a') X and $\emptyset$ are closed; (b') the arbitrary intersection of closed sets is closed; (c') a finite union of closed sets is closed.

Let A be a subset of a topological space. The smallest closed set containing A is called the *closure* $\bar{A}$ of A. To emphasize the topology in which the closure is taken, $\bar{A}$ will be denoted by $Cl_u A$. The following statements about the closure are immediate: For each subset A of X_u, (a") $A \subset \bar{A}$, (b") $\bar{\bar{A}} = \bar{A}$, (c") for any finite number of subsets A_i $(1 \le i \le n)$,

$$\overline{\bigcup_{1 \le i \le n} A_i} = \bigcup_{i=1}^{n} \bar{A_i}, \text{(d")} \; \bar{X} = X \text{ and } \bar{\emptyset} = \emptyset.$$ A subset A of X_u is closed if and only if $\bar{A} = A$.

Let A and B be two subsets of a topological space X_u. A is said to be *dense* in B if $\bar{A} \supset B$. A topological space X_u is said to be *separable* if X_u contains a countable dense subset.

Let A be any subset of a topological space X_u. Then A can also be topologized as follows: For each u-open set U in X_u, define $U \cap A$ to be open in A. Then it is easy to check that the family $\{U \cap A\}$, when U runs over u, defines a topology on A. This topology on A is called the *induced* or *relative* topology of A.

2. METRIC SPACES

An important subclass of topological spaces is the class of metric spaces.

Let E be a set. Suppose there exists a real valued function d defined on the ordered pairs (x, y), $x, y \in E$, satisfying the following axioms:

(m_1) $d(x, y) \ge 0,$ for all $x, y \in E$

(m_2) $d(x, y) = 0$ if and only if $x = y$

(m_3) $d(x, y) = d(y, x)$

(m_4) $d(x, y) + d(y, z) \ge d(x, z),$ $x, y, z \in E.$

Then E is said to be a *metric space*. For each positive $r > 0$ and a fixed $x_0 \in E$, the subset $B_r(x_0) = \{x \in E : d(x, x_0) < r\}$ is called on *open ball* of radius r. The subset $\{x \in E : d(x, x_0) \leq r\}$ is called a *closed ball*. Let $u = \{U\}$ be the family of subsets of a metric space E such that, for each U, if $x \in U$ then there exists $r > 0$ such that $B_r(x) \subset U$. Then it is easy to check that the family u defines a topology on E and E is called a *metric topological* space or simply a *metric space*. It is immediate that each open ball of a metric space is an open set and similarly each closed ball is a closed set.

Examples. (1) Let $E = R$, the real line and $d(x, y) = |x - y|$. Then R is a metric space and so a topological space. Apart from this metric topology, indeed, there are other topologies on R as well, e.g., the discrete and the indiscrete.

(2) $E = R^n$, the space of n-tuples; i.e., $x = (x_1, \ldots, x_n)$ where $x_i \in R$, $1 \leq i \leq n$. The metric d is defined by the following formula:

$$d(x, y) = \sqrt{\sum_{i=1}^{n} |x_i - y_i|^2},$$

$x = (x_1, \ldots, x_n)$ and $y = (y_1, \ldots, y_n)$. Then E is a metric space. It is called the Euclidean n-dimensional space. If $n = 1$, then (2) and (1) coincide.

(3) Let I denote the closed unit interval $[0, 1]$. Let $E = C(I)$ denote the set of all continuous real-valued functions on I. Define d as follows:

$$d(f, g) = \sup_{0 \leq x \leq 1} |f(x) - g(x)|,$$

in which f and g are continuous functions on I. Then E is a metric space.

If the topology of a given topological space X can be described by a metric then X is said to be *metrizable*.

Indeed, as shown above, every arbitrary set can be topologized by at least two topologies, viz., the discrete and the indiscrete. The question of whether a topological space is metrizable is not that easy. Of course, an indiscrete space is not metrizable, as will appear easily from the separation axioms dealt with in the sequel. But a discrete space is metrizable and the metric is the following: For each $x, y \in X$, define

$$d(x, y) = \begin{cases} 1 \text{ if } x \neq y \\ 0 \text{ if } x = y. \end{cases}$$

Then it is easy to check that d defines a metric that describes the discrete topology. Sometimes this metric is called a *trivial metric*.

A sequence $\{x_n\}$ in a metric space is said to be a *Cauchy sequence* if for each $\varepsilon > 0$ there exists a positive integer n_0 depending upon ε such that $d(x_n, x_m) < \varepsilon$ for all $n, m \geq n_0$. A sequence $\{x_n\}$ in E is said to *converge* to a point $x_0 \in E$ if for each $\varepsilon > 0$ there exist $n_0 = n_0(\varepsilon)$ such that $d(x_n, x_0) < \varepsilon$ for all $n \geq n_0$.

It is easy to show that if a sequence converges then it is a Cauchy sequence. But the converse is not true in general. If each Cauchy sequence converges in a metric space E, then E is said to be a *complete* metric space. The reader can verify that the metric spaces mentioned in examples (1) to (3) above are all complete metric spaces.

If a metric space E is not complete, then by a well-known procedure it can be completed to a complete space $\hat{E}$ so that E forms a dense subset of $\hat{E}$. $\hat{E}$ is called the *completion* of E and is nothing more than the set of all equivalence classes of Cauchy sequences in E. A metric space E is complete if, and only if, it coincides with its completion $\hat{E}$. (Observe that we do not distinguish between the two sets if their elements can be put in a $1:1$ correspondence.)

Observe that the notion of completion depends upon that of Cauchy sequences. Since the latter notion is not necessarily defined on a nonmetric topological space, the notion of completion need not be defined for a non-metric topological space either. We shall see in the sequel that there is a generalization (uniform spaces) of metric spaces on which the completion can be defined.

In connection with metric spaces the following notions are useful:

A subset A of a metric space E is said to be *nowhere-dense* or *nondense* if $(\bar{A})^0 = \varnothing$ (i.e., if the interior of the closure of A is empty). A countable union of nondense sets is said to be of the *first category*. A set that is not of the first category is of the *second category*. A topological space of the second category is also known as a *Baire space*.

Theorem 1. (*Baire.*) *Every complete metric space E is of the second category, or is a Baire space.*

PROOF. Suppose $E = \bigcup_{n \geq 1} A_n$, where each A_n is nondense. For each n, $E \sim \bar{A}_n$ is open and everywhere-dense because otherwise there would be an open ball consisting of only points of $\bar{A}_n$, thus contradicting the fact that the interior of $\bar{A}_n$ is empty. Since each A_n is nondense, there exists a sequence $\{\varepsilon_n\}$ of positive real numbers and a sequence $\{x_n\}$ of elements in E such that $\bar{B}_{\varepsilon_n}(x_n) \cap A_n = \varnothing$ for each n. We may assume (by induction) that $\varepsilon_{n+1} < \varepsilon_n$, $\varepsilon_n \to 0$ ($\{\varepsilon_n\}$ converges to zero) and $B_{\varepsilon_n}(x_n) \supset \bar{B}_{\varepsilon_{n+1}}(x_{n+1})$ for each $n \geq 1$. It is easy to see that by this choice $\{x_n\}$ is a Cauchy sequence and, hence, converges to some $x_0 \in E$ because the latter is complete by hypothesis. Since $x_m \in B_n(x_n) \subset \bar{B}_n(x_n)$ for all $m \geq n$, for a fixed n, $x_0 \in \bar{B}_n(x_n)$ for each n. Since $\bar{B}_n(x_n) \cap A_n = \varnothing$ for each n, $x_0 \notin A_n$ for each $n \geq 1$, i.e., $x_0 \notin \bigcup_{n \geq 1} A_n = E$, which is a contradiction. Hence, E is of the second category.

It is easy to see from the above theorem that every nonempty open subset of a complete metric space is of the second category.

A subset A in a topological space E is said to be *residual* if $E \sim A$ is of the first category.

From Theorem 1 it follows that all residual subsets in a complete metric space are nonempty.

3. NEIGHBORHOOD SYSTEMS

Let X_u be a topological space. Let $x \in X$. A subset P of X is said to be a *u-neighborhood* of x if there exists a *u*-open set U such that $x \in U \subset P$. Observe that a neighborhood of a point $x \in X$ is not necessarily an open set. However, it is quite clear that an open set is a neighborhood of each point contained in it. The following connection between the open sets and neighborhoods is easy to verify: A subset A of X is *u*-open if, and only if, for each $x \in A$ there exists a neighborhood P_x of x such that $P_x \subset A$.

For each $x \in X_u$, let $\mathscr{U}_x$ denote the totality of all *u*-neighborhoods of x. Then the following properties are immediately established by using the definitions of neighborhoods and open sets:

(n_1) For each member U_x in $\mathscr{U}_x$, $x \in U_x$.

(n_2) If U_x is in $\mathscr{U}_x$ and W is any subset of X such that $U_x \subset W$, then W is in $\mathscr{U}_x$.

(n_3) Each finite intersection of sets in $\mathscr{U}_x$ is also in $\mathscr{U}_x$.

(n_4) If U_x is in $\mathscr{U}_x$, then there exists a V_x in $\mathscr{U}_x$ such that $V_x \subset U_x$ and $U_x \in \mathscr{U}_y$ for each $y \in V_x$, where $\mathscr{U}_y$ is the totality of all *u*-neighborhoods of y. We prove the following:

Proposition 1. *Let X be a set. Suppose for each $x \in X$, there exists a system $\mathscr{U}_x$ of subsets of X satisfying the above conditions (n_1)–(n_4). Then there exists a unique topology u on X such that $\mathscr{U}_x$ is precisely the system of all u-neighborhoods of x for each $x \in X$.*

PROOF. Let u denote the collection of subsets consisting of $\varnothing$ and of all U such that $U \in \mathscr{U}_x$ whenever $x \in U$. Then we show that u defines a topology on X. Since for each $x \in X$, $X \in \mathscr{U}_x$ owing to (n_2), u contains X and by definition of u, $\varnothing$ is in u. Let $\{U_i\}$ ($1 \leq i \leq n$) be a finite family of sets in u. Then for each $x \in \bigcap_{i=1}^{n} U_i$, $x \in U_i$ for all i, and, hence, $U_i \in \mathscr{U}_x$ for all i. But then by (n_3), $\bigcap_{i=1}^{n} U_i \in u$. Finally, let U_α be an arbitrary family of sets in u. Let $x \in \bigcup_\alpha U_\alpha$. Then $x \in U_\alpha$ for some α and, hence, $U_\alpha \in \mathscr{U}_x$. But $\bigcup_\alpha U_\alpha \supset U_\alpha$ and so $\bigcup_\alpha U_\alpha \in u$ by (n_2). Hence, u defines a topology on X and, for each x, $U_x \in \mathscr{U}_x$ is a *u*-neighborhood of x. By using (n_4) one sees easily that $\mathscr{U}_x$ is precisely the system of all *u*-neighborhoods of x for each $x \in X$.

If one discusses filter-bases instead of filters, (§6), then condition (n_2) is dropped. Thus Proposition 1 may be stated as well for filter-bases $\mathscr{B}_x$ for each x.

From Proposition 1 combined with the foregoing remarks, it follows that a topology on a set can be defined either by its open sets or by neighborhoods of each point. There are other ways of defining topologies, e.g., by assigning closures or limit points of sets (see overleaf). It is interesting to note that all these processes are practically equivalent in the sense that, given one process, one can define others in terms of the given one.

Let X_u be a topological space and A a subset of X. An element $x \in X$ is said to be a *limit point* of A if each neighborhood P_x of x meets A, i.e., $A \cap P_x \neq \emptyset$. One knows that a subset A together with all its limit points coincides with the closure $\bar{A}$ of A. Thus, the following statements are equivalent: (a) A is closed; (b) $A = \bar{A}$; (c) each limit point of A is in A; (d) no point of $E \sim A$ is a limit point of A. If the space X_u is metric then $x \in \bar{A}$ if, and only if, $d(x, A) = \inf_{y \in A} d(x, y) = 0$, which is equivalent to: There exists a sequence $\{x_n\}$ in A such that $\{x_n\}$ converges to x.

4. BASES AND SUBBASES

Let X_u be a topological space with $u = \{U\}$ as its family of open sets. A subfamily $\{U_\alpha\}$ of u is said to be a *base* of u if for each $x \in U \in \{U\}$ there exists an α such that $x \in U_\alpha \subset U$. Or, equivalently, each U in $\{U\}$ is the union of members of $\{U_\alpha\}$. A subfamily $\{U_\beta\}$ of $\{U\}$ is said to form a *subbase* for u if the family of finite intersections of members of $\{U_\beta\}$ forms a base of u. Clearly the set of open balls of a metric space forms a base of the metric topology. In particular, the set of open bounded intervals on the real line forms a base of its metric topology. The sets of open intervals $\{x \in R : x < a\}$ or $\{x \in R : x > a\}$, when a runs over the real line R, form subbases of the metric topology.

By localizing the above considerations at a single point of a topological space, one can define bases and subbases of the neighborhood systems of points. More specifically, let $\mathscr{U}_x$ denote the totality of all u-neighborhoods of $x \in X_u$. Then a subfamily $\mathscr{V}_x$ of $\mathscr{U}_x$ is said to be a *base* of $\mathscr{U}_x$ or to form a *fundamental system of neighborhoods* of x if for each U_x in $\mathscr{U}_x$, there exists a V_x in $\mathscr{V}_x$ such that $V_x \subset U_x$. A subfamily $\mathscr{W}_x$ of $\mathscr{U}_x$ is said to be a *subbase* of $\mathscr{U}_x$ if the family of finite intersections of members of $\mathscr{W}_x$ forms a base for $\mathscr{U}_x$. Clearly, in a metric space X_u, the system $\{B_{1/n}(x)\}$ (for integer $n \geq 1$) of open balls forms a base of neighborhoods of x. If for each $x \in X_u$ there exists a countable base of the neighborhood system at x, X_u is said to satisfy the *first axiom of countability*. A topological space in which the system of open sets has a countable base is said to satisfy the *second axiom of countability*. Obviously, every topological space satisfying the second axiom of countability also satisfies the first axiom of countability. But the converse is not true, e.g., the real line with the discrete topology. As indicated above, every metric space satisfies the first axiom of countability.

Let X_u be a topological space satisfying the second axiom of countability. If $\{U_n\}$ is a countable base of u, then by selecting distinct elements, one from each U_n, there is produced a countable dense set. Thus X_u is separable. In particular, a metric space satisfying the second axiom of countability is separable. If a metric space E is separable, let $\{x_n\}$ be a countable dense subset of E and for each x_n consider $\{B_r(x_n)\}$, where r runs over the positive rationals. Then it is easy to check that $\{B_r(x_n)\}$, r positive rational, $n \geq 1$, forms a countable base of the metric topology. In other words, a metric space is separable if, and only if, it satisfies the second axiom of countability.

5. SEPARATION AXIOMS IN TOPOLOGICAL SPACES

Let X_u be a topological space. The following separation axioms in X_u are known as Alexandroff and Hopf separation axioms:

A_0: Of each couple x, $y \in X$, $x \neq y$, there is at least one in an open neighborhood excluding the other.

A_1: Of each couple x, $y \in X$, $x \neq y$, each has an open neighborhood not containing the other.

A_2: Of each couple x, $y \in X$, $x \neq y$, both have open neighborhoods U, V such that $x \in U$, $y \in V$ and $U \cap V = \varnothing$ (i.e., U and V are *disjoint*).

A_3: Let C be any closed subset of X_u and x any point of X such that $x \notin C$. Then there exist disjoint open sets U and V such that $x \in U$, $C \subset V$.

A_4: Let C and x be as in A_3. Then there exists a real valued continuous function on X to $[0, 1]$ such that $f(x) = 0$ and $f(C) = 1$.

A_5: Let C_1 and C_2 be any two disjoint closed subsets of X. Then there exist two disjoint open subsets U_1 and U_2 such that $C_i \subset U_i$, $i = 1, 2$.

A topological space satisfying A_i is called a T_i-*space* ($i = 0, \ldots, 5$). A T_2-space is also known as a *Hausdorff* space. In the literature, there is a little confusion in labeling these spaces. Therefore, the reader is advised to adopt our usage for this book.

Clearly, A_i implies A_{i-1} for $i = 1, 2$. Moreover, A_4 implies A_3. By Urysohn's lemma it follows that A_5 is equivalent to: For each pair of disjoint closed subsets $C_i (i = 1, 2)$ there exists a continuous real valued function f on X to $[0, 1]$ such that $f(C_1) = 0$, $f(C_2) = 1$. Thus, if each point or singleton in a topological space is closed, then A_5 implies A_4 and also A_3 implies A_2. Furthermore, it is known that each point of a topological space X is closed if, and only if, X is a T_1-space. Thus, for T_1-spaces, A_i implies A_{i-1} for $i = 1, 2, 3, 4, 5$.

A T_1-space satisfying $A_i (i = 3, 4, 5)$ will be called *regular, completely regular*, and *normal*, respectively.

As in cases of T_i-spaces ($i = 1, 5$) (see the paragraph before the last), T_3-spaces X can also be characterized as follows: For each $x \in X$ and each open neighborhood U of x there exists an open neighborhood V of x such that $\bar{V} \subset U$.

There are examples of T_i-spaces that are not T_{i+1}-spaces ($i = 0, \ldots, 4$). As we shall see in Chapter III, axioms A_0–A_4 are equivalent for T_0-topological groups, and axioms A_0–A_5 are equivalent for locally compact T_0-groups. Note that an indiscrete space is *not* a T_0-space and hence not a T_i-space for $i = 1, \ldots, 5$. We also have the following:

Proposition 2. *A T_3-space (in particular, a regular space) satisfying the second axiom of countability is a T_5-space (normal).*

PROOF. Let E be a T_3-space and let $\{U_i\}$ $(i \geq 1)$ denote a countable base of the topology of E. Let C_1, C_2 be two disjoint closed subsets of E. For each $x \in C_1$, there exist disjoint open sets V_x and U_x such that $x \in U_x$, $C_2 \subset V_x$. One can choose U_x from $\{U_i\}$. Thus, as x runs over C_1, only a countable number of elements $U_{x_i}(i \geq 1)$ from $\{U_i\}$ cover C_1. To each U_{x_i} there corresponds a V_{x_i} containing C_2. Now let $U = \bigcup_{n\geq 1}\left\{U_{x_n} \sim \bigcup_{i=1}^{n-1} \bar{V}_{x_i}\right\}$

and $V = \bigcup_{n\geq 1}\left\{V_{x_n} \sim \bigcup_{i=1}^{n-1} \bar{U}_{x_i}\right\}$. Then it is easy to check that U contains C_1

$\left(\text{because } C_1 \subset \bigcup_{n\geq 1} U_{x_n} = \bigcup_{n\geq 1}\left\{U_{x_n} \sim \bigcup_{i=1}^{n-1} \bar{V}_{xi}\right\}\right)$ and $C_2 \subset V$ (because each V_x

contains C_2). Further, it is quite obvious that U and V are disjoint open sets.

One sees that the full force of the hypothesis, viz., the second countable axiom, has not been used. What one requires is the countable covering axiom. To make things precise, let us consider the following:

Let X be a set. A family $\mathscr{F} = \{P_\alpha\}$ of subsets in X is said to be a *covering* of X if for each $x \in X$ there exists a member in $\mathscr{F}$ that contains x. If X_u is a topological space then a covering $\mathscr{F} = \{P_\alpha\}$ is said to be *open* if each P_α is an open set. A topological space is said to be *Lindelöf* if every covering is reducible to a countable open covering. Clearly, a topological space satisfying the second axiom of countability is Lindelöf. But the converse is not true, e.g., let R be the real line with the topology having the system of half-open and half-closed intervals $\{[a, b): a \leq x < b, a, b \in R\}$ as its base. Then R is a Lindelöf space but does not satisfy the second axiom of countability.

Now one sees from the proof of Proposition 2, that one can show the following:

Proposition. 3   *Every Lindelöf T_3-space (or regular) is a T_5-space (or normal).*

On the other extreme, every metric or metrizable space is normal. However, a normal space need not be metrizable. But by a well-known metrization theorem it follows that a regular space satisfying the second axiom of countability is metrizable (Kelley,[27] p. 125).

For results concerning the question whether the subspaces of T_i-spaces are T_i-spaces, one is again referred to the standard books on topology, e.g., Kelley[27] or Bourbaki.[4] We note, however:

Proposition 4.   *Let X_α $(\alpha \in A)$ be a family of T_5-spaces (or normal spaces) such that $X_\alpha \cap X_\beta = \varnothing$, $\alpha \neq \beta$ for all α, $\beta \in A$. Let $X = \bigcup_{\alpha \in A} X_\alpha$ be endowed with a topology defined as follows: A subset U of X is open if, and only if, $U \cap X_\alpha$ is open for each α. Then X is a T_5-space (or normal).*

PROOF. Let C_1 and C_2 be two disjoint closed subsets of X. Then for each α, $C_1 \cap X_\alpha$ and $C_2 \cap X_\alpha$ are disjoint and closed subsets of X_α. Hence, normality of X_α implies there are two disjoint open sets U_α and V_α in X_α such that $C_1 \cap X_\alpha \subset U_\alpha$ and $C_2 \cap X_\alpha \subset V_\alpha$. Now let $U = \bigcup\limits_{\alpha \in A} U_\alpha$ and $V = \bigcup\limits_{\alpha \in A} V_\alpha$, then U and V are open by the definition of the topology, and disjoint because $X_\alpha \cap X_\beta = \varnothing$, $\alpha \neq \beta$, and $U_\alpha \cap V_\alpha = \varnothing$. Clearly, $C_1 \subset U$ and $C_2 \subset V$. This completes the proof.

6. NETS AND FILTERS

Let A be a nonempty set such that there is a binary relation "$\geq$" defined for the elements of A satisfying the following conditions:

(i) "$\geq$" is reflexive, i.e., $a \geq a$ for each $a \in A$.

(ii) "$\geq$" is transitive, i.e., if $a \geq b$ and $b \geq c$ then $a \geq c$, a, b, $c \in A$.

(iii) If a, $b \in A$ then there exists $c \in A$ such that $c \geq a$ and $c \geq b$.

Then $(A, \geq)$ or simply A is said to be a *directed* set.

Let X be any set. A subset $\{x_\alpha\}(\alpha \in A)$ of X is said to be a *net* if A is a directed set of indices. A net $\{x_\alpha\}(\alpha \in A)$ in a topological space X_u *converges* to $x \in X$ if for each neighborhood U of x there exists a $\beta \in A$ such that for all $\alpha \geq \beta$, $x_\alpha \in U$.

One immediately observes that the concept of a directed set is a generalization of the positive integers with their natural ordering, and that of a net is the generalization of a sequence. Furthermore, the convergence of nets in a general topological space replaces the concept of the convergence of sequences in metric spaces. As was remarked before, limits of sequences give rise to the closures of sets and thereby determine the topology of metric spaces. Similarly, nets in the general topological spaces lead to the same goal by following a similar path. We remark that although the limits of convergent sequences in metric spaces are unique, the same may not be true for topological spaces. For example, every net in an indiscrete space converges to every point of the space. However, if a topological space is Hausdorff then the limits of convergent nets are unique, and conversely.

Appropriately imitating the concept of subsequences one can define *subnets* as well. Apart from the theory of convergence by means of nets, there is another theory of convergence due to H. Cartan,[8] which is followed up by Bourbaki and others of the French school of mathematics. It is based on the concept of a filter as defined below:

Let X be a given set and $\mathscr{F} = \{F_\alpha\}$ a nonempty family of subsets of X. $\mathscr{F}$ is said to be a *filter* on X if the following conditions hold:

(F_1) Any subset of X containing any F_α is in $\mathscr{F}$.

(F_2) The intersection of any finite number of F_α's is also in $\mathscr{F}$.

(F_3) The empty set $\varnothing$ does not belong to $\mathscr{F}$.

A nonempty subfamily $\mathscr{G} = \{G_\beta\}$ of a filter $\mathscr{F} = \{F_\alpha\}$ on X is called a *base* of $\mathscr{F}$ if

(i) The intersection of any two members in $\mathscr{G}$ contains a member of $\mathscr{G}$.

(ii) Each F_α contains a G_β.

A nonempty family $\mathscr{F} = \{F_\alpha\}$ on X satisfying (i) and not including $\varnothing$ is said to be a *filter-base*.

Let X_u be a topological space and $\mathscr{F}$ a filter on X. $\mathscr{F}$ is said to converge to $x \in X$ if for each neighborhood U of x there exists an F_α such that $F_\alpha \subset U$. The following facts are easily deduced:

(a) A subset U of X is open if, and only if, U is a member of each filter that converges to a point of U.

(b) If a filter $\mathscr{F}$ converges to x_0, then every other filter $\mathscr{K}$ that is finer than $\mathscr{F}$ (i.e., every member of $\mathscr{K}$ is contained in some member of $\mathscr{F}$) also converges to x_0.

Taking all filters and ordering them by inclusion, one can obtain a partially ordered system of filters. By Zorn's lemma (Kelley[27]), one gets a maximal filter, which is called an *ultrafilter*. In a topological space, an ultrafilter either converges or has no limit point.

One knows that if $\{x_\alpha\}(\alpha \in A)$ is a net then the family of all subsets $F_\alpha = \{x_\beta : \beta \geq \alpha,\ \alpha,\ \beta \in A\}$ is a filter. If A is a countable set, then the set of tails of a sequence forms a *countable* filter, which is sometimes called an *elementary filter* associated with the sequence. It is also well-known that given a filter one can obtain a net. Thus, the two theories of convergence are essentially the same.

7. COMPACT, LOCALLY COMPACT AND CONNECTED SPACES

A topological space X_u is said to be *compact* if each open covering (§5) of X_u is reducible to a finite open covering. As one knows from general topology, for a Hausdorff space X_u, the following statements are equivalent:

(a) X_u is compact.

(b) Let $\{C_\alpha\}$ be a family of closed sets in X_u. If the intersection of every finite subfamily of C_α's is nonempty (such a system is said to satisfy the *finite-intersection property*) then $\bigcap_\alpha C_\alpha \neq \varnothing$. In other words, if $\bigcap_\alpha C_\alpha = \varnothing$ then there exists a finite subfamily $\{C_{\alpha_i}\}(1 \leq i \leq n)$ such that $\bigcap_{i=1}^{n} C_{\alpha_i} = \varnothing$.

(c) Every ultrafilter on X_u is convergent.

(d) Each filter on X_u has at least one limit point in X.

A subset Y of a topological space X_u is a *compact set* if Y, endowed with the relative topology, is compact. A finite union of compact sets is compact. A closed subset of a compact space is compact. A compact subset of a Hausdorff topological space is closed. In a nonHausdorff space a compact subset need not be closed, e.g., in an indiscrete space a compact subset is not necessarily closed. Every compact Hausdorff space is normal. Every compact discrete space is finite.

A subset Y of a topological space X_u is said to be *relatively compact* if $\overline{Y}$ is a compact set. Every subset of a compact space is relatively compact.

A topological space X_u is said to be *locally compact* if for each $x \in X_u$ there exists an open neighborhood U of x such that $\bar{U}$ is compact. A

Hausdorff topological space is locally compact if, and only if, each point has a compact neighborhood. A closed subset of a locally compact Hausdorff space is locally compact. Using the same arguments as those used in Theorem 1, and replacing balls by compact neighborhoods, one sees that every Hausdorff locally compact topological space is a Baire space. Every Hausdorff locally compact topological space is completely regular and hence regular.

Proposition 5. *A locally compact Hausdorff topological space E is normal if it is the union of an increasing sequence $\{U_n\}$ of open sets such that $\bar{U}_n$ is compact for each n.*

PROOF. By hypothesis $\bar{U}_n \subset \bar{U}_{n+1}$ and $\bar{U}_{n+1} \sim U_n = \bar{U}_{n+1} \cap (E \sim U_n)$ is compact. Furthermore, $E = \bigcup_{n \geq 1} U_n = \bigcup_{n \geq 1} \bar{U}_n = \bigcup_{n \geq 0} (\bar{U}_{n+1} \sim U_n)$, where $U_0 = \varnothing$. Let $\{P_\alpha\}$ be any open covering of E. Since each $\bar{U}_{n+1} \sim U_n$ is compact, a finite subcovering of $\{P_\alpha\}$ covers $\bar{U}_{n+1} \sim U_n$ for each n. Since the countable union of finite families of sets is a countable family, this shows that a countable number of members of $\{P_\alpha\}$ cover E. This proves that E is a Lindelöf space, and hence Proposition 3, §5, applies.

A noncompact topological space X_u can be compactified as follows: Let X_u be such a space and let "ω" be an abstract element not in X_u. Let $\tilde{X} = X \cup \{\omega\}$. Define the topology on $\tilde{X}$ as the one containing all the open sets of X_u and those sets $\tilde{U}$ containing ω, whose complement $\tilde{X} \sim \tilde{U}$ is compact in X_u. Then it is easy to check that $\tilde{X}$ is compact. $\tilde{X}$ is Hausdorff if, and only if, X is a Hausdorff locally compact topological space. $\tilde{X}$ is called the *one-point compactification* of X. The one-point compactification is useful in deriving the hereditary properties of a space from its one-point compactification. There are, indeed, other compactifications, e.g., the Stone-Čech compactification. But we shall not make use of them in this book. For greater detail see Kelley.[27]

A topological space X_u is said to be *connected* if there exist no nonempty disjoint open sets P and Q such that $P \cup Q = X$. In other words, a topological space is connected if, and only if, there are no other closed and open subsets of X except X and the empty set $\varnothing$. The Euclidean spaces $R^n (n \geq 1)$ are connected. The set of rational numbers with the relative topology is not connected.

A subset of a topological space is said to be *connected* if it is a connected space in the relative topology. The maximal connected subset containing a point of a topological space is called the *component* of that point.

If A is a connected set, so is $\bar{A}$. Hence, every component is closed. If $X_\alpha (\alpha \in A)$ is a family of connected sets such that $\bigcap_{\alpha \in A} X_\alpha \neq \varnothing$, then $\bigcup_{\alpha \in A} X_\alpha$ is also connected.

A topological space is said to be *locally connected* if each open neighborhood of every point contains a connected open neighborhood. Each component of an open subset of a locally connected space is open.

8. MAPPINGS

Let E and F be two sets. A law or correspondence that assigns to each element of E at most one element of F is called a *mapping, function* or *map* from E to F. If f denotes a mapping then, for each $x \in E$, $f(x)$ denotes the *image* of x in F. If, for each pair x_1, $x_2 \in E$, $x_1 \neq x_2$ implies $f(x_1) \neq f(x_2)$, then f is said to be *one-to-one* or 1:1. If for each element y of F there exists $x \in E$ such that $f(x) = y$, then f is said to be *onto*. For each subset A of E, $f(A) = \{y \in F : y = f(x), x \in A\}$. The set of all mappings of E into F is denoted by F^E. If f is a 1:1 mapping of E onto F, then there exists an inverse mapping $f^{-1}: F \to E$ defined by $f^{-1}(y) = x$ if, and only if, $f(x) = y$, where $x \in E$, $y \in F$.

Let E and F be two topological spaces. A mapping f of E into F is said to be *continuous* if for each open set V in F, $f^{-1}(V) = \{x \in E : f(x) \in V\}$ is an open set of E. The following statements are equivalent:

(a) f is continuous.

(b) For each closed set C in F, $f^{-1}(C)$ is closed in E.

(c) For each member V of a subbase of the topology in F, $f^{-1}(V)$ is open in E.

(d) For each $x \in E$ and each neighborhood V of $f(x)$, $f^{-1}(V)$ is a neighborhood of x in E.

(e) For each $x \in E$ and each neighborhood V of $f(x)$ there exists a neighborhood U of x such that $f(U) \subset V$.

(f) For each subset B of F, $\overline{f^{-1}(B)} \subset f^{-1}(\bar{B})$.

(g) For each subset A of E, $f(\bar{A}) \subset \overline{f(A)}$.

For the proof of these equivalences see Kelley,[27] p. 86.

Let f be a continuous mapping of a topological space E into another topological space F. Let G be a subset of E. Since the intersection of each open set in E with G is an open set in G by definition, the *restriction* $f \,|\, G$ of f on G is also continuous. Furthermore, if f is a continuous mapping of E into F and if g is a continuous mapping of F into another topological space H, then the composition mapping $g \circ f$ of E into H, defined by $g \circ f(x) = g(f(x)) \in H$ for each $x \in E$, is also continuous.

A notion weaker than that of continuity is the following: A mapping f of a topological space E into another topological space F is *almost continuous* at $x \in E$ if for each neighborhood V of $f(x)$, $\overline{f^{-1}(V)}$ is a neighborhood of $x \in E$. f is said to be almost continuous everywhere if it is so at each $x \in E$.

Since $\overline{f^{-1}(V)} \supset f^{-1}(V)$, it follows that if f is continuous then it is almost continuous. But the converse is not true, e.g., let $E = F = R$, the real line. Define

$$f(x) = \begin{cases} 1 \text{ if } x \text{ is rational} \\ 0 \text{ if } x \text{ is irrational} \end{cases}$$

Then it is quite clear that f is not continuous at any $x \in R$. In other words,

f is totally discontinuous. However, for $\varepsilon > 0$, the sets

$$U_1 = \{x \in E : |f(x) - 1| < \varepsilon\}$$

and

$$U_2 = \{x \in E : |f(x)| < \varepsilon\}$$

are everywhere dense in the real line and, therefore, their closures coincide with R, which is a neighborhood of each of its points. This shows that f is almost continuous.

By using a result attributable to H. Blumberg,[3] it is possible to show that every real function on the real line has points of almost continuity everywhere dense in the real line (Husain[24]). One can actually extend this for any real valued function on the metric Baire space (Husain[24]) by using a result, obtained by Bradford and Goffman,[7] which is an extension of Blumberg's result referred to just above. However, we shall not use these results in this book.

Another notion weaker than that of continuity is described with the help of the following: Let E and F be two sets. $E \times F$ denotes the set of all pairs (x, y), $x \in E$, $y \in F$, and is called the product of E and F. Let f be a mapping of E into F. The subset $\{(x, f(x)) : x \in E\}$ of $E \times F$ is called the *graph* of f in $E \times F$.

If E and F are topological spaces one can define a topology on $E \times F$ as follows: A subset U of $E \times F$ is open if, and only if, for each $(x, y) \in U$ we have $U \supset U_1 \times U_2$, where U_1 is an open neighborhood of x in E and U_2 an open neighborhood of y in F. In other words, each U is the union of sets of the type $U_1 \times U_2$, U_1 open in E and U_2 open in F. Now we have the following:

Proposition 6. *Let f be a continuous mapping of a topological space E into a Hausdorff topological space F. Then the graph of f is a closed subset of $E \times F$. But the converse is not true.*

PROOF. Let $G = \{(x, f(x)) : x \in E\}$ be the graph of f in $E \times F$. Let $(x, y) \notin G$. Then $f(x) \neq y$. Since F is Hausdorff, there exist disjoint open sets V_1 and V_2 such that $y \in V_1$ and $f(x) \in V_2$. Since f is continuous, there exists an open neighborhood U of x such that $f(U) \subset V_2$. Since $V_1 \cap V_2 = \varnothing$, $f(U) \cap V_1 = \varnothing$. Clearly, $U \times V_1$ is a neighborhood of (x, y). If $G \cap (U \times V_1) \neq \varnothing$, then there exists $x_1 \in U$ such that $f(x_1) \in V_1$, which is impossible because $f(U)$ and V_1 are disjoint. Therefore, (x, y) has an open neighborhood completely contained in the complement of G. Since (x, y) is arbitrary, we see that G is closed.

For the converse, let $E = R_u$ and $F = R_d$, where R is the real line, u the usual metric topology and d the discrete topology. Let f be the identity mapping of E onto F. Since f is $1:1$, the graph of f is the same as that of the inverse $f^{-1} : F \rightarrow E$, which is continuous (because $d \supset u$). Hence, the graph of f^{-1} is closed and so is the graph of f. But f cannot be continuous because otherwise d must be equal to u, which is not true.

A mapping f of a topological space E into a topological space F is said to be *open* (or *closed*) if for each open (or closed) set U of E, $f(U)$ is open (or closed). One knows from the general topology that a continuous mapping is neither necessarily open nor closed, etc. However, a $1:1$ open mapping of a topological space E onto another topological space F has a closed graph. A notion weaker than that of openness is the following: A mapping of E into F is *almost open* at $x \in E$ if for each open neighborhood U of x, $\overline{f(U)}$ is a neighborhood of $f(x)$. f is almost open everywhere whenever it is almost open at each $x \in E$. Clearly, an open mapping is almost open. But the converse need not be true. A $1:1$ continuous and open mapping of a topological space E onto another topological space is called a *homeomorphism*. If there is a homeomorphism between two topological spaces then they are called *homeomorphic* with each other.

A continuous image of a compact set is compact. Moreover, if f is a $1:1$ continuous mapping of a compact space onto a Hausdorff topological space, then f is open and therefore a homeomorphism. Since a subset of the set of real numbers is compact if, and only if, it is closed and bounded, it follows that every real-valued continuous function on a compact space is bounded. Moreover, since the infimum and supremum of a closed bounded subset of the real line are in that subset, every real-valued continuous function on a compact space assumes its minimum and maximum values.

It is instructive to note that none of the results of the previous paragraph is true for locally compact spaces.

A continuous image of a connected space is connected.

9. DIRECT PRODUCTS

Let $X_\alpha(\alpha \in A)$ be a family of sets. $X = \prod_{\alpha \in A} X_\alpha$ denotes the set of all mappings $x:A \to X$ such that $x(\alpha) = x_\alpha \in X_\alpha$ for each $\alpha \in A$. X is called the *direct product* or a *Cartesian product* or simply a *product* of X_α's. Each x_α is called the *projection* of x on X_α. The mapping $p_\alpha:x \to x_\alpha$ of X onto X_α is called the αth *projection mapping*. If $X_\alpha = F$ for each $\alpha \in A$, then X is denoted by F^A.

If $\{X_\alpha\}(\alpha \in A)$ is a family of topological spaces then X can be topologized as follows: Let $\{U_\alpha\}$ denote the family of open sets in X_α for each $\alpha \in A$. Then the family $\{p_\alpha^{-1}(U_\alpha)\}$ forms a subbase of a topology called the *product topology* on X, and X, endowed with the product topology, is called a *topological product*. If there is no confusion, a topological product will also be called a product.

By the definition of the product topology on X, it is clear that the projection mapping $p_\alpha:X \to X_\alpha$ is continuous and open for each $\alpha \in A$. Let f be a mapping of a topological space E into a topological product $X = \prod_{\alpha \in A} X_\alpha$. Then f is continuous if, and only if, $p_\alpha \circ f$ is continuous for each $\alpha \in A$.

A net in $X = \prod_{\alpha \in A} X_\alpha$ converges to a point $x \in X$ if, and only if, its projection converges to $p_\alpha(x)$ for each $\alpha \in A$. This result justifies the term *coordinatewise convergence* often associated with the product topology. In case each $X_\alpha = F$ for $\alpha \in A$ then $\prod_{\alpha \in A} X_\alpha = F^A$ and a net $(f_\beta, \beta \in B)$ in F^A converges to a point $f \in F^A$ if, and only if, for each $a \in A$, $\{f_\beta(a): \beta \in B\}$ converges to $f(a)$ in F. This justifies the label of *pointwise* or *simple convergence* in the product space.

The following few results concerning the products are well-known:
(a) $X = \prod_{\alpha \in A} X_\alpha$ is a Hausdorff topological space if, and only if, each X_α is a Hausdorff space. (b) X satisfies the first axiom of countability if each X_α does and if A is a countable set of indices. (c) X is metric if each X_α is metric and if A is countable. Suppose $\{X_n\}$ is a sequence of metric spaces with metrics d_n. Then for $x, y \in X = \prod_{n=1}^{\infty} X_n$,

$$d(x, y) = \sum_{n=1}^{\infty} 2^{-n} \frac{d_n(x_n, y_n)}{1 + d_n(x_n, y_n)}$$

is a metric on X, where $x_n = p_n(x)$ and $y_n = p_n(y)$.
A very important result in this connection is the following:

Theorem 2. (*Tychonoff.*) $X = \prod_{\alpha \in A} X_\alpha$ *is compact if, and only if, each* X_α *is compact.*

See Kelley[27] for the proof.
Theorem 2 is not true for locally compact spaces. However, if each X_α is locally compact and all X_α's except a finite number of them are compact, then the product is also locally compact.

10. UNIFORM SPACES AND ASCOLI'S THEOREM

Let X be a set and $X \times X = X^2$ the product of X by itself. Let V denote a collection of elements (x, y) of X^2. Then V^{-1} denotes the set of all pairs (y, x) such that $(x, y) \in V$. If U and V are any two subsets of X^2 then UV denotes the set of all pairs (x, z) such that for some $y \in X$, $(x, y) \in U$ and $(y, z) \in V$. By replacing V by U in UV one can define U^2. The set $\{(x, x): x \in X\}$ is called the *diagonal.*

A set X is said to be a *uniform space* if there exists a filter $\mathcal{K} = \{U\}$ on X^2 satisfying the following properties:
(a^0) Each U in $\mathcal{K}$ contains the diagonal.
(b^0) If U is in $\mathcal{K}$ then U^{-1} is also in $\mathcal{K}$.
(c^0) For each U in $\mathcal{K}$, there exists a V in $\mathcal{K}$ such that $V^2 \subset U$.
The filter $\mathcal{K}$ satisfying (a^0)–(c^0) is called a *uniformity* for the set X. One sees easily that properties (a^0)–(c^0) are generalizations of metric properties

(see §2). Clearly a metric determines a uniformity. A metric space with the uniformity determined by its metric becomes a uniform space.

As a metric on a set X defines a (metric) topology so does the uniformity in the following way: Let $\mathscr{K}$ be a uniformity for a set X. For each $x \in X$, let

$$U_x = \{y \in X : (x, y) \in U\}.$$

Then as U runs over $\mathscr{K}$, the system U_x is a base of the neighborhood system at x under a topology. A set T in X is open in the topology of the uniform space if, and only if, for each $x \in T$, there exists a U in $\mathscr{K}$ such that $U_x \subset T$. Whenever we talk of a uniform space without explicitly mentioning its topology in the sequel, we shall mean that the uniform space is endowed with this topology. A uniform space is a Hausdorff space if, and only if, $\bigcap U$ is the diagonal. Every Hausdorff uniform space is completely regular (Kelley,[27] p. 188).

A net $\{x_\alpha\}(\alpha \in \Gamma)$ in a uniform space X is a *Cauchy net* if for a given U in the uniformity $\mathscr{K}$ of X there exists $\alpha_0 \in \Gamma$ such that for all $\alpha, \beta \geq \alpha_0$ $(\alpha, \beta \in \Gamma)$, $(x_\alpha, x_\beta) \in U$. A net $\{x_\alpha\}(\alpha \in \Gamma)$ converges to $x_0 \in X$ if for each U in $\mathscr{K}$ there exists an $\alpha_0 \in \Gamma$ such that for all $\alpha \geq \alpha_0$, $(x_\alpha, x_0) \in U$.

Thus, as in the case of metric spaces, one can define completeness of uniform spaces. If a uniform space is not complete, then it can be completed by means of Cauchy nets. Indeed, one can deal with Cauchy filters instead of nets when the necessity arises. A filter $\mathscr{F} = \{F_\alpha\}$ on X is a *Cauchy filter* if for each U in $\mathscr{K}$ (the uniformity on X), there exists an F_α such that $(x, y) \in U$ for all $x, y \in F_\alpha$. In a similar fashion one defines the convergence of filters in a uniform space. A subset of a uniform space is also a uniform space under the relative uniformity defined as usual by taking intersections.

Proposition 7. *A closed subset F of a complete uniform space E is complete. Also, a complete subset M of a Hausdorff uniform space is closed.*

PROOF. Let $\{x_\alpha\}(\alpha \in \Gamma)$ be a Cauchy net in F. Then $\{x_\alpha\}(\alpha \in \Gamma)$ is a Cauchy net in E and therefore converges to $x_0 \in E$. Since x_0 is a limit point of $\{x_\alpha\}(\alpha \in \Gamma)$ in F and since the latter is closed, it follows that $x_0 \in F$.

For the other part, let $x \in \bar{M}$. Then there exists a net $\{x_\alpha\}(\alpha \in \Gamma)$ in M which converges to x. Clearly $\{x_\alpha\}(\alpha \in \Gamma)$ is a Cauchy net and hence converges in M because M is complete. This shows $x \in M$ and hence M is closed.

By using a characterization of compact spaces (§7) it follows that a compact uniform space is complete.

Let $X_\alpha(\alpha \in A)$ be a uniform space for each α and let $X = \prod_{\alpha \in A} X_\alpha$. Then one can define the *product uniformity* for X as follows: Let $\mathscr{K}_\alpha = \{U_\alpha\}$ denote the uniformity for X_α for each α and let

$$U = \{(x, y) \in X^2 : (p_\alpha(x), p_\alpha(y)) \in U_\alpha\},$$

where p_α is the projection of X onto X_α. Then U forms a subbase for the product uniformity. In other words, the family of finite intersections of U's is a base for the product uniformity. The product uniformity describes precisely the product topology on X (Kelley,[27] p. 183). If each X_α is complete so is X.

Let E and F be two uniform spaces with uniformities $\mathscr{K} = \{U\}$ and $\mathscr{G} = \{V\}$, respectively. A mapping f of E into F is said to be *uniformly continuous* relative to $\mathscr{K}$ and $\mathscr{G}$ if for each $V \in \mathscr{G}$, the set $\{(x, y) \in E^2 : (f(x), f(y)) \in V\}$ is a member of $\mathscr{K}$. A uniformly continuous mapping is always continuous, but the converse is not true, e.g., $f(x) = x^2$ is continuous on the real line but not uniformly continuous. However, every continuous mapping of a compact uniform space into another uniform space is uniformly continuous.

By specializing the product $X = \prod_{\alpha \in A} X_\alpha$ of uniform spaces when $X_\alpha = F$ for each $\alpha \in A$, we see that F^A is a uniform space and the pointwise convergence topology is precisely the product topology. Let f be a mapping of a topological space E into a uniform space F, then $f \in F^E$. Let $C(E, F)$ denote the set of all continuous mappings of E into F. Then $C(E, F) \subset F^E$. We can endow $C(E, F)$ with the relative pointwise convergence topology induced from F^E.

Let E and F be any two topological spaces. Let $\mathfrak{S}$ be any collection of subsets of E and $\{V\}$ a base of the open sets in F. Let

$$T(M, V) = \{f \in F^E : f(M) \subset V\}.$$

Then the family $\{T(M, V)\}$, when M runs over $\mathfrak{S}$ and V over $\{V\}$, forms a subbase of a topology called an $\mathfrak{S}$-topology on F^E. If F is a Hausdorff space and $\mathfrak{S}$ is a covering of E, the $\mathfrak{S}$-topology is also Hausdorff. There are two important special cases of the $\mathfrak{S}$-topology: (a) $\mathfrak{S}$ is the collection of all finite subsets of E. In this case, the $\mathfrak{S}$-topology is nothing but the pointwise convergence topology and is denoted by p. (b) $\mathfrak{S}$ is the collection of all compact subsets of E. In this case, the $\mathfrak{S}$-topology is called the *compact-open topology* and denoted by k. Since every finite subset is a compact subset, $k \supset p$. Therefore, any subset G of F^E which is k-compact is also p-compact. But the converse is not true.

Our main interest in the remainder of this section is devoted to finding out the conditions under which a subset G of $C(E, F) \subset F^E$ is k-compact. First we define the following: A family G of functions in $C(E, F)$, where E is a topological space and F a uniform space, is said to be *equicontinuous* at $x \in E$ if for each member V in the uniformity $\mathscr{G}$ of F, there exists a neighborhood U of x such that

$$(f(y), f(x)) \in V$$

for all $y \in U$ and $f \in G$.

The p-closure of an equicontinuous set in F^E is also equicontinuous (Kelley,[27] p. 232). On each equicontinuous set in $C(E, F)$, the relative topology k coincides with the relative topology p.

Now we have the following:

Theorem 3. (*Ascoli.*) *Let E be a Hausdorff locally compact topological space and F a Hausdorff uniform space. Let $C(E, F)$ be endowed with the k-topology. Then a subset G of $C(E, F)$ is k-compact if, and only if:*
(a) *G is k-closed.*
(b) *The closure of the set $\{f(x) : f \in G\}$, for each $x \in E$, is compact in F.*
(c) *G is equicontinuous.*

PROOF. For the "if" part, let G be a subset of $C(E, F)$ satisfying (a)–(c). Then by (b), Cl_pG is p-compact. Since G is equicontinuous by (c) Cl_pG is also equicontinuous as remarked above. Hence Cl_pG is k-compact because k coincides with p on equicontinuous sets. Thus, $Cl_kG = G$ being a closed subset of a k-compact subset Cl_pG (since $p \subset k$ implies $Cl_kG \subset Cl_pG$), it follows that G is k-compact.

For the "only if" part, suppose G is a k-compact subset of $C(E, F)$. Then clearly G is k-closed because k is a Hausdorff topology. Since $k \supset p$ and G is k-compact, G is p-compact. Thus, (a) and (b) are satisfied. To establish (c), let $x_0 \in E$ and let V_1 and V be two members of the uniformity of F such that $V_1^3 \subset V$. Since each $f \in G$ is continuous at x_0, there exists a compact (because E is locally compact) neighborhood U_f of x_0 such that $(f(x), f(x_0)) \in V_1$ for all $x \in U_f$. The family $\{T(U_f, V_1)\}_{f \in G}$, where $T(U_f, V_1) = \{g \in C(E, F) : (g(x), g(x_0)) \in V_1 \text{ for all } x \in U_f\}$, is a k-open covering of G. Since G is k-compact by assumption, a finite subfamily $\{T(U_i, V_1)\}$, $(1 \leq i \leq n$, for some n), covers G. Setting $U = \bigcap_{i=1}^{n} U_i$, we see easily that for all $f \in G$ and for all $x \in U$, $(f(x), f(x_0)) \in V_1^3 \subset V$. Since V is arbitrary, it proves that G is equicontinuous; and, thus, having established (c), we have proved the theorem.

11. GROUPS AND LINEAR SPACES

(a) Groups
A nonempty set G with the law of composition "$\circ$" among its elements is called a group if the following conditions hold:
(i) For $a, b \in G$, $a \circ b \in G$.
(ii) The composition is associative, i.e., $a \circ (b \circ c) = (a \circ b) \circ c$.
(iii) There exists a *unit element* or *identity* e in G, i.e., $e \circ a = a \circ e = a$ for each $a \in G$.
(iv) Each $a \in G$ has its *inverse* a^{-1}, i.e., $a^{-1} \circ a = a \circ a^{-1} = e$.
We will suppress the notation "$\circ$" of composition altogether. Instead we shall use the usual multiplication "ab" or addition "$a + b$" as convenience suggests. A group with a law of multiplication (or addition) will

be called a multiplicative (or additive) group. A group G is said to be *commutative* or *abelian* if $ab = ba$ (or $a + b = b + a$) for each a, $b \in G$. The identity of an additive group will be denoted by 0 instead of e and the inverse by $-a$ instead of a^{-1}.

Indeed, the set of all integers is an additive abelian group with 0 as its identity. The sets of all rational and real numbers also form additive abelian groups. The sets of all nonzero rational and real numbers form multiplicative abelian groups. Furthermore, the set of all complex numbers is an additive abelian group and the set of all nonzero complex numbers is a multiplicative abelian group. Also the set of all complex numbers z such that $|z| = 1$($|z|$ denotes the modulus of z) is a multiplicative abelian group. There are nonabelian groups as well, e.g., the set of $n \times n$ nonsingular matrices forms a nonabelian multiplicative group.

A subset H of a group G is said to be a *subgroup* if for each a, $b \in H$, $ab^{-1} \in H$. In other words: H contains the identity of G; for each a, $b \in H$, $ab \in H$; and for each $a \in H$, $a^{-1} \in H$. Every group contains at least two subgroups, viz., the whole group G and the group consisting of the identity only. These two subgroups are said to be *trivial subgroups*. For any subgroup H of a group G, $H^2 = H H = H$.

In any group G, if a, $b \in G$ then $(ab)^{-1} = b^{-1}a^{-1}$. Further, if H is any subgroup of G then for any $a \in G$, $a^{-1}Ha$ is also a subgroup of G. If $a \in H$ then trivially $a^{-1}Ha = H$ or $Ha = aH$. However, if $a \notin H$ and $a \in G$, then $a^{-1}Ha$ need not coincide with H. Whenever $a^{-1}Ha = H$ for each $a \in G$ then H is said to be an *invariant, normal* or *distinguished* subgroup of G. We shall use only the first term in the sequel whereas the term "normal" will be reserved for a concept in topology. In an abelian group, each subgroup is invariant.

For any subgroup H of a group G and any element $a \in G$, aH (or Ha) is called a *left* (or *right*) *coset*. For any two right cosets Ha and Hb either $Ha \cap Hb = \varnothing$ or $Ha = Hb$ (if, and only if, $ab^{-1} \in H$). To show this, one sees that $ab^{-1} \in H$ if, and only if, $ba^{-1} \in H$, since H is a subgroup. Thus $ab^{-1} \in H$ implies $Hab^{-1} \subset H^2 = H$. Hence, $Ha \subset Hb$. Similarly, $ba^{-1} \in H$ implies $Hb \subset Ha$. Thus $Ha = Hb$. The other implication is trivial. One can show that $G = \bigcup aH = \bigcup Ha$, where the union is taken over all pairwise disjoint cosets. For finite groups G this decomposition of a group into pairwise disjoint cosets gives the so-called Lagrange's theorem which states that the order of a subgroup H of G divides the order of G, i.e., ord $(G) = k$ ord (H), when ord denotes the order and k a positive integer (Van der Waerden).[45]

Let H be an invariant subgroup of a group G. Then $aH = Ha$ for each $a \in G$. That means the left cosets are identically equal to the right cosets. We may define multiplication on the set of all cosets $\{aH\}$ as follows: $aH \, bH = a(Hb)H = abH^2 = abH$. Furthermore, for each aH, $(aH)H = aH^2 = aH$ and $H(aH) = aH^2 = aH$. Hence, H plays the role of the identity. Moreover, for each aH, $(a^{-1}H)(aH) = a^{-1}(Ha)H = H^2 = H$ shows that

$a^{-1}H$ is the inverse of aH. Hence, the totality of all cosets $\{aH\}$ forms a group. This is called the *quotient* group and denoted by G/H.

Let G be a group. The set C of all elements $a \in G$ that commute with each $x \in G$ forms a group. For, if $a, b \in C$ then $ax = xa$ and $bx = xb$ for all $x \in G$. Hence, $(ab)x = a(bx) = a(xb) = (ax)b = x(ab)$. Therefore, $ab \in C$. The associative law is hereditary. Clearly, the identity e of G is in C. Furthermore, for each $a \in C$, $ax = xa \Leftrightarrow x^{-1}a^{-1} = a^{-1}x^{-1}$. Since $G = G^{-1}$, $a^{-1} \in C$. Hence, C is a group. Moreover, $ax = xa$ for all $a \in C$ and $x \in G$ shows that C is an invariant subgroup of G. C is called the *center* of G. If G is abelian, then $G = C$. Since $e \in C$, $C \neq \varnothing$.

Let G and H be two groups. A mapping f of G into H is said to be a *homomorphism* if for each pair $x, y \in G$, $f(xy) = f(x)f(y)$ in the multiplicative case, or $f(x + y) = f(x) + f(y)$ in the additive case. (Observe that from our expressions it appears that a homomorphism is defined when G and H are either both multiplicative or both additive. But this need not be the case. This symbolism is adopted for the sake of convenience.) If $G = H$ then f is said to be an *endomorphism* of G. A $1:1$ and onto homomorphism is said to be an *isomorphism*. An isomorphism of G onto itself is called an *automorphism*. If f is an isomorphism of G onto H, then G and H are said to be *isomorphic*. Since $ex = x$, $f(x) = f(ex) = f(e)f(x)$. This implies that a homomorphism always maps the identity of G onto the identity of H. Furthermore, for each $x \in G$, $x\,x^{-1} = e$ implies $f(x)\,f(x^{-1}) = f(e) = e' \in H$, or $f(x^{-1}) = [f(x)]^{-1}$. Thus, a homomorphism maps inverses into inverses.

The set $\{x \in G : f(x) = e' \in H\}$, where e' is the identity of H, is easily checked to be an invariant subgroup of G. It is denoted by $f^{-1}(e')$ and called the *kernel* of the homomorphism f. The image of a group G under a homomorphism f is also a subgroup of H.

Let G be a group and H an invariant subgroup of G. For each $x \in G$, let xH denote the unique coset to which x belongs in the decomposition of G into pairwise disjoint cosets. Let $\varphi(x) = xH$ define a mapping of G into G/H. Then it is easy to check that φ is a homomorphism of G onto G/H. φ is called the *natural* or *canonical* homomorphism of G onto G/H.

In connection with homomorphisms of groups there are two important theorems, viz.:

Theorem 4. *Let G be a group, H an invariant subgroup of G, and M any subgroup of G. Then H is an invariant subgroup of HM, which is a subgroup of G, and $H \cap M$ is an invariant subgroup of M. Furthermore, the groups HM/H, $\varphi(M)$ and $M/(M \cap H)$ are isomorphic.*

Theorem 5. *Let G and H be any two groups and let f be a homomorphism of G onto H. Let M be an invariant subgroup of H, $f^{-1}(M) = N$, and $f^{-1}(e') = Q$. Then the groups G/N, H/M and $(G/Q)/(N/Q)$ are isomorphic.*

For proofs see Van der Waerden,[45] pp. 141–142.

(b) Linear spaces

Let E be an abelian additive group and K the field of real or complex numbers. For each $x \in E$ and $\lambda \in K$, let λx be an element in E. Assume the following conditions are satisfied:

(i) $\lambda(x + y) = \lambda x + \lambda y$

(ii) $(\lambda + \mu)x = \lambda x + \mu x$, $\lambda, \mu \in K$

(iii) $\lambda(\mu x) = (\lambda \mu)x$

(iv) $1x = x$, $x \in E$ and $1 \in K$.

Then E is called a *vector space* or *linear space* over the field K.

A subset F of a vector space E is said to be a *subspace* if F is a linear space over the same field with addition and scalar multiplication induced from E. Let F be a subspace of a vector space E; the quotient space E/F is defined in the same way as that for a group.

Let E be a vector space over the field K, and let $\{x_i\}(1 \leq i \leq n)$ be a finite subset of E. For $\lambda_i \in K$, $1 \leq i \leq n$, the object $\sum_{i=1}^{n} \lambda_i x_i$ is called a *linear combination* of the x_i. A subset F of E is said to be *linearly independent* if F is not empty or $\{0\}$ and no element of F is a linear combination of any finite subset of other elements in F. A maximal linearly independent subset of a vector space is said to be a *basis*.

Every vector space has a basis. Any two bases of a vector space have the same cardinal number. The cardinal number of a basis of a vector space is called its *dimension*. If the cardinal number of a basis of a vector space is that of a finite subset of positive integers, the vector space is called *finite-dimensional*.

A vector space E over the field K is said to be an *algebra* if the following conditions hold: For each pair $x, y \in E$ the product xy is defined and belongs to E. The product is associative and the following distributive laws hold: $x(y + z) = xy + xz$, $(x + y)z = xz + yz$, $\lambda(xy) = (\lambda x)y = x(\lambda y)$ for all $x, y, z \in E$ and $\lambda \in K$. If, in addition, $xy = yx$ for all $x, y \in E$ then the algebra is said to be *commutative*.

The sets of all complex numbers, all quaternions as well as the set of all $n \times n$ real matrices form algebras over the field of real numbers.

Let E be a real or complex vector space. A real valued function $\| \cdot \|$ on E is said to be a *norm* if

(i) $\|x\| \geq 0$ for all $x \in E$

(ii) $\|x\| = 0$ if, and only if, $x = 0$

(iii) $\|\lambda x\| = |\lambda| \|x\|$, λ scalar

(iv) $\|x + y\| \leq \|x\| + \|y\|$.

E with a norm is said to be a *normed space*. Defining $d(x, y) = \|x - y\|$, one sees that E is a metric space with d as its metric induced by the norm. A normed space complete in the metric induced by the norm is called a *Banach space*.

An algebra E that is a normed (Banach) space and in which the inequality $\|xy\| \leq \|x\| \|y\|$, for all $x, y \in E$, holds, is called a *normed (Banach) algebra*.

An important subclass of the class of all normed (Banach) spaces is the class of *scalar-product* or *pre-Hilbert* (*Hilbert*) spaces.

A normed space E is said to be a *pre-Hilbert* space if its norm satisfies the following additional property:

(v) $\|x + y\|^2 + \|x - y\|^2 = 2(\|x\|^2 + \|y\|^2)$ for all $x, y \in E$.

The reason why a pre-Hilbert space E is called a scalar product is that by means of (v) one can define a scalar product function $\langle x, y \rangle$ on $E \times E$, by the following formula:

$$\langle x, y \rangle = \tfrac{1}{4} [\|x + y\|^2 - \|x - y\|^2$$
$$+ i \|x + iy\|^2 - i \|x - iy\|^2], \text{ where } i = \sqrt{-1}.$$

It can be shown (although the calculations are a little tedious) that the scalar product function satisfies the following familiar properties:

(a) $\langle x_1 + x_2, y \rangle = \langle x_1, y \rangle + \langle x_2, y \rangle$

(b) $\langle \lambda x, y \rangle = \lambda \langle x, y \rangle$

(c) $\langle x, y \rangle = \overline{\langle y, x \rangle}$

(d) $\langle x, x \rangle \geq 0$

(e) $\langle x, x \rangle = 0$ if, and only if, $x = 0$.

From the definition of the scalar product, it follows that $\|x\|^2 = \langle x, x \rangle$ for each $x \in E$, or $\|x\| = +\sqrt{\langle x, x \rangle}$.

A complete pre-Hilbert is thus a Hilbert space.

A subset F of a normed (or pre-Hilbert) space E is called a *subspace* of E if F is a vector subspace of E and if F is endowed with the norm induced from E. Similarly, one defines *subalgebras* of a normed algebra. It is clear that a closed subspace (or subalgebra) of a Banach space (or Banach algebra) is itself a Banach space (or Banach algebra), since a closed subset of a complete metric space is complete.

Let E be a normed space and F a closed subspace of E . The set of all cosets $x + F$ is defined as the *quotient space* and it is denoted by E/F. E/F is a normed space with the norm

$$\|x + F\| = \inf_{y \in F} \|x + y\|.$$

If E is a Banach (or Hilbert) space, then E/F is also a Banach (or Hilbert) space. If E is a Hilbert space, we see that $E/F = F^\perp$ (see below).

Let M be any subset of a Hilbert space E. The set $M^\perp = \{x \in E : \langle x, y \rangle = 0$ for all $y \in M\}$ is clearly a closed subspace of E and $M^\perp$ is called the *orthogonal complement* of M.

The following is a well-known result in the theory of Hilbert spaces (e.g., see Halmos[17]): Let M be a closed subspace of a Hilbert space E. Then each $x \in E$ can be written uniquely as the sum of elements in M and $M^\perp$, viz., $x = x_1 + x_2$, where $x_1 \in M$, $x_2 \in M^\perp$.

Let E be a vector space over the field K. A mapping f of E into K is said to be a *functional*. A functional is said to be *linear* if $f(\lambda x + \mu y) = \lambda f(x) + \mu f(y)$ for $x, y \in E$ and $\lambda, \mu \in K$.

Let E be a normed space. A linear functional f on E is said to be *bounded* if there exists $M > 0$ such that $|f(x)| \leq M \|x\|$ for all $x \in E$. Every bounded linear functional on E is continuous, and conversely. The least M for which $|f(x)| \leq M \|x\|$ for all $x \in E$, is called the *norm* of f and is denoted by $\|f\|$.

Let F be a subspace of a normed space E, and let f be a bounded linear functional on F. Then there exists a bounded linear functional g on E such that $g = f$ on F and $\|g\| = \|f\|$. This result is the *Hahn-Banach theorem* (cf., Husain,[22] pp. 14–17).

A very useful relation between continuous linear functionals on a Hilbert space and the scalar product function is given by the following:

Theorem 6. (*Riesz Representation Theorem.*) *Let E be a Hilbert space. For each linear continuous functional f on E there exists a unique element $y_0 \in E$ such that*

$$f(x) = \langle x, y_0 \rangle$$

for all $x \in E$.

PROOF. If $f = 0$, we take $y_0 = 0$ and the theorem is proved. If $f \neq 0$, the set $M = \{x \in E : f(x) = 0\}$ is a proper (i.e., $M \neq E$) closed linear subspace of E. Let $z \in M^\perp$, $z \neq 0$. Let α be the scalar such that if $y_0 = \alpha z$ then $f(y_0) = \langle y_0, y_0 \rangle$. (Clearly $\bar{\alpha} = f(z)/\langle z, z \rangle$.) It is easy to check that E/M is one-dimensional. Hence every $x \in E$ has the unique representation as the sum $x_1 + x_2$, where $x_1 \in M$ and $x_2 \in M^\perp = E/M$. Since E/M is one-dimensional, and since $y_0 \in M^\perp$, for all x_2 in $M^\perp$, $x_2 = \lambda y_0$. Hence, $x = x_1 + \lambda y_0$. But then

$$f(x) = f(x_1 + \lambda y_0) = f(x_1) + \lambda f(y_0) = \lambda f(y_0)$$
$$= \lambda \langle y_0, y_0 \rangle = \langle \lambda y_0, y_0 \rangle = \langle x_1 + \lambda y_0, y_0 \rangle$$
$$= \langle x, y_0 \rangle$$

for all $x \in E$. This completes the proof.

Let E be a Banach space. The set E' of all continuous linear functionals of E is said to be the *dual* of E. For $x \in E$ and $x' \in E'$, $\langle x, x' \rangle$ denotes the *value* of x' at x. The set E' can be endowed with several topologies. One topology is $\sigma(E', E)$, the coarsest topology in E' with respect to which the mapping $x' \to \langle x, x' \rangle$ is continuous; $\sigma(E', E)$ is called the weak or weak* topology on E'. The other is the norm topology defined by the norm: $\|x'\| = \sup_{\|x\| \leq 1} |\langle x, x' \rangle|$. The set E', endowed with the norm topology, is also a Banach space. Thus, one defines the dual E'' of E'. The set E'' is called the *bidual* of E. It is known that E can be identified with a subspace of E''.

It is clear that if $x \in E$ and $x' \in E'$, then we obtain

$$|\langle x, x' \rangle| \leq \|x'\| \|x\|,$$

(*Hölder's inequality*).

If E is a Hilbert space, by theorem 6, we can identify E' with E and thus from Hölder's inequality we obtain *Schwarz's inequality*, viz., for $x, y \in E$,

$$|\langle x, y \rangle| \leq \|x\| \, \|y\|,$$

where $\langle x, y \rangle$ is the scalar product function.

If $\{H_\alpha\}$ is a family of mutually orthogonal (i.e., $\langle x_\alpha, x_\beta \rangle = 0$ for all $x_\alpha \in H_\alpha$ and $x_\beta \in H_\beta$, $\alpha \neq \beta$) closed subspaces of a Hilbert space H, then for any $x \in H$,

$$(1) \qquad\qquad \sum_\alpha \|x_\alpha\|^2 \leq \|x\|^2,$$

in which x_α is the projection of x on H_α and the sum on the left is taken in the generalized sense, i.e., via the filter generated by the family of finite subsets of the index set $\{\alpha\}$. (1) is known as the *Bessel inequality*. If $\{H_\alpha\}$ spans H, i.e., the closure of the direct sum $\sum_\alpha H_\alpha$ coincides with H, then $\sum_\alpha \|x_\alpha\|^2 = \|x\|^2$ (the *Parséval equality*).

Let E be a normed space and E' its dual. If a subspace M of E is dense in E then for $x' \in E'$, $\langle x, x' \rangle = 0$ for all $x \in M$ implies $x' = 0$.

We shall come across a number of examples of Banach spaces and Banach algebras in the sequel. One of the most important algebras is the following function algebra.

Let X be a locally compact Hausdorff space. A complex-valued function f on X is said to *vanish at infinity* if for each $\varepsilon > 0$ there exists a compact subset C of X such that $|f(x)| \leq \varepsilon$ for all $x \in X \sim C$. Let $C_\infty(X)$ denote the set of all continuous and complex-valued functions defined on X and vanishing at infinity. It is easy to see that under the pointwise addition, scalar multiplication and multiplication (i.e., if $f, g \in C_\infty(X)$ then $(f + g)(x) = f(x) + g(x)$, $\lambda f(x) = \lambda[f(x)]$, and $fg(x) = f(x)g(x)$, for $x \in X$ and λ complex), $C_\infty(X)$ is an algebra. $C_\infty(X)$ is a subalgebra of $C(X)$, the set of all continuous bounded complex-valued functions on X. If $\|f\|_\infty = \sup_{x \in X} |f(x)|$ denotes the norm on $C(X)$, then it can be shown that $C(X)$ is a Banach algebra and $C_\infty(X)$ is a closed subalgebra of $C(X)$. Hence $C_\infty(X)$ is also a Banach algebra.

If X is compact, then it is clear that $C_\infty(X) = C(X)$. If X is a noncompact locally compact Hausdorff space, let X_∞ denote the one-point compactification. Each $f \in C_\infty(X)$ can be extended uniquely to an element $\tilde{f} \in C(X_\infty)$ by defining $\tilde{f} = f$ on X and $\tilde{f}(\infty) = 0$. Thus, $C_\infty(X)$ can be identified with a subalgebra of $C(X_\infty)$.

A subalgebra A of $C(X)$ is said to be *closed under the lattice operations* if for $f, g \in A$, the functions $f \vee g = h$ and $f \wedge g = k$ are in A, where h and k are defined by $h(x) = \max(f(x), g(x))$ and $k(x) = \min(f(x), g(x))$ for each

$x \in X$. It is easy to see that

$$f \vee g = 2^{-1}(f + g + |f - g|)$$

and

$$f \wedge g = 2^{-1}(f + g - |f - g|)$$

where $|f|$ is the usual absolute value function, viz., $|f|(x) = |f(x)|$. If A is closed under the lattice operations and if $f_i \in A$, $(1 \le i \le n)$, then $f_1 \wedge f_2 \wedge \cdots \wedge f_n$ and $f_1 \vee f_2 \vee \cdots \vee f_n$ are in A.

A very useful theorem that we shall need in the sequel is the following:

Theorem 7. (*Stone-Weierstrass Theorem.*) *Let X be a locally compact Hausdorff space. Let A be a subalgebra of $C_\infty(X)$ such that*
(a) *for $x \ne y$, x, $y \in X$ there exists $f \in A$ such that $f(x) \ne f(y)$;*
(b) *for each $x \in X$ there exists $f \in A$ such that $f(x) \ne 0$;*
(c) *for each $f \in A$, $\bar{f} \in A$, where $f(x) = \overline{f(x)}$ (complex conjugate).*
Then A is dense in the Banach algebra $C_\infty(X)$, when the latter is endowed with the sup norm topology. In other words, if A is a norm closed subalgebra of $C_\infty(X)$ satisfying $(a) - (c)$ then $A = C_\infty(X)$.

PROOF. Considering the one-point compactification X_∞ of X and identifying $C_\infty(X)$ with $C(X_\infty)$ (see the above discussion), it is sufficient to prove the theorem for compact X.

Let $C_R(X)$ denote the set of all real continuous functions on a compact Hausdorff space X. First, we show that if A is a norm closed subalgebra of $C_R(X)$ satisfying (a) and (b), then $A = C_R(X)$. We break the proof in several steps.

(i) *If $f \in A$ then $|f| \in A$.* For, if t is a real variable, $|t|$ is clearly a continuous real function. Hence, by the classical Weierstrass theorem, for each $\varepsilon > 0$, there exists a polynomial p_1 such that $|\,|t| - p_1(t)| < \varepsilon/2$ for all t, $|t| \le \|f\|$. Hence, if $p = p_1 - p_1(0)$ then $|\,|t| - p(t)| < \varepsilon$ for all t, $|t| \le \|f\|$. Since A is an algebra and p a polynomial, $p(f) \in A$ and $|\,|f(x)| - p(f(x))| < \varepsilon$ for each $x \in X$. Whence, $\|\,|f| - p(f)\| \le \varepsilon$. Thus, $|f| \in A$, since $|f|$ is a limit point of A and A is closed.

(ii) *A is closed under the lattice operations.* For, if $f, g \in A$ then, in view of (i)

$$f \vee g = 2^{-1}(f + g + |f - g|) \in A$$

and

$$f \wedge g = 2^{-1}(f + g - |f - g|) \in A$$

because A is an algebra.

(iii) *If $x, y \in X$, $x \ne y$ then for any real numbers r and s, there exists a continuous real function f in A such that $f(x) = r$, $f(y) = s$.* Indeed, by (a) there exists $g \in C_R(X)$ such that $g(x) \ne g(y)$. If

$$f(z) = r \frac{g(z) - g(y)}{g(x) - g(y)} + s \frac{g(z) - g(x)}{g(x) - g(y)}, \quad z \in X$$

then $f \in C_R(X)$ and $f(x) = r$, $f(y) = s$. Since A is an algebra, $f \in A$.

(iv) To show that $A = C_R(X)$, let $f \in C_R(X)$ and let $\varepsilon > 0$. For $x \neq y$, $x, y \in X$, by (iii) there exists $f_{x,y} \in A$ such that $f_{x,y}(x) = f(x)$ and $f_{x,y}(y) = f(y)$. Let $U_{x,y} = \{z \in X : f_{x,y}(z) < f(z) + \varepsilon\}$ and $V_{x,y} = \{z \in X : f_{x,y}(z) > f(z) - \varepsilon\}$. Clearly, $U_{x,y}$ and $V_{x,y}$ are open sets. For fixed y and variable x, $\{U_{x,y}\}$ is an open covering of X. Since X is compact, only a finite subfamily $\{U_{x_i,y}\}$, $(1 \leq i \leq n)$, covers X. Let $f_{x_i,y}$ be the corresponding function which defines $U_{x_i,y}$. Then $f_y = f_{x_1,y} \wedge f_{x_2,y} \wedge \cdots \wedge f_{x_n,y} \in A$ by (ii), $f_y(x) = \wedge f_{x_i,y}(x) = f(x)$, $f_y(z) < f(z) + \varepsilon$ for all $z \in X$, and $f_y(z) > f(z) - \varepsilon$ for $z \in \bigcap_{i=1}^{n} V_{x_i,y} = V_y$. Now varying y, similarly we obtain a finite family $\{f_{y_i}\}$, $(1 \leq i \leq m)$ of functions in A. Putting $f_\varepsilon = f_{y_1} \vee f_{y_2} \vee \cdots \vee f_{y_m}$, we see that $f_\varepsilon \in A$ by (ii), and $f - \varepsilon < f_\varepsilon < f + \varepsilon$ on X. Since ε is arbitrary, $f \in \bar{A} = A$. This completes the proof of the theorem in the real case.

For the complex case, we first observe that the set A_R of all real functions in A is a norm-closed subalgebra of $C_R(X)$, since A is closed in $C(X)$ by hypothesis. If we show that $A_R = C_R(X)$, then it is easily seen that $A = C(X)$. But to show that $A_R = C_R(X)$, we make use of condition (c). For any $f \in C(X)$, $f = \mathscr{R}(f) + i\mathscr{I}(f)$, $i = \sqrt{-1}$, where $\mathscr{R}(f)$ and $\mathscr{I}(f)$ are real continuous functions. Hence $\mathscr{R}(f) = 2^{-1}(f + \bar{f}) \in A_R$ and $\mathscr{I}(f) = (2i)^{-1}(f - \bar{f}) \in A_R$, because of condition (c), $\bar{f} \in A$ and because A_R is an algebra. Now if $x \neq y$, $x, y \in X$, by (a) there exists $f \in A$ such that $f(x) \neq f(y)$. This implies that the real part $\mathscr{R}(f)$ of f satisfies the conditions: $\mathscr{R}(f)(x) \neq \mathscr{R}(f)(y)$, and $\mathscr{R}(f) \in A_R$. Hence, A_R satisfies (a). Suppose $f \in A$ such that $f(x) \neq 0$ then $f\bar{f} = |f|^2 \in A_R$ and $|f|^2(x) \neq 0$. Hence, A_R satisfies (b). Therefore, by the real case proved above, $A_R = C_R(X)$. This completes the proof of the theorem.

II

Semitopological Groups

In this chapter we make a study of semitopological groups—a notion weaker than the well-known one of topological groups. A topological group is always a semitopological group, but the converse is not true as shown by an example. We derive here conditions that imply a semitopological group is a topological group.

12. THE CONCEPT OF A SEMITOPOLOGICAL GROUP

Definition 1. (a) A topological space G that is also a group is called a *semitopological group* if the mapping

$$g_1:(x, y) \rightarrow xy$$

of $G \times G$ onto G is continuous in *each* variable separately.

(b) A topological space G that is also a group is called a *topological group* if the mapping g_1 is continuous in *both* variables together and if the inversion mapping

$$g_2:x \rightarrow x^{-1}$$

of G onto G is also continuous.

If the group operation is addition instead of multiplication, xy and x^{-1} should be regarded as $x + y$ and $-x$, respectively. The identity of a multiplicative group will be denoted by e and that of an additive group by 0.

Proposition 1. *Every topological group is a semitopological group. But the converse is not true.*

27

PROOF. The first statement is clearly true. To show that the converse is not true, let $G = R$, the real line as an additive abelian group. Let G be endowed with a topology which has $\{[a, b): -\infty < a < x < b < \infty\}$, the system of left closed and right open intervals as its base. Since for each neighborhood $[a, b)$ of the identity 0, $\left[a, \dfrac{b}{2}\right)$ is also a neighborhood of 0, it follows that the mapping g_1 is continuous in both variables together at 0. It is easily seen that g_1 is continuous everywhere. Hence, G is a semitopological group. However, the mapping $g_2 : x \to -x$ is not continuous at 0 because if $[0, b)$ is a neighborhood of 0, then there is no neighborhood V of 0 such that $-V \subset [0, b)$. Therefore, G is not a topological group. This completes the proof.

If we put $UV = \{xy : x \in U, y \in V\}$ and $U^{-1} = \{x^{-1} : x \in U\}$, where U and V are subsets of a group G, and in the additive case, $U + V = \{x + y : x \in U, y \in V\}$, $-U = \{-x : x \in U\}$, then the continuity of the mappings g_1 and g_2 can be expressed as follows:

g_1 is continuous in x (or y) if, and only if, for each neighborhood W of xy there exists a neighborhood U (or V) of x (or y) such that $Uy \subset W$ (or $xV \subset W$). If G is abelian, then the right and left continuities of $(x, y) \to xy$ in each variable are equivalent.

Moreover, g_1 is continuous in both x and y if, and only if, for each neighborhood W of xy there exist a neighborhood U of x and a neighborhood V of y such that $UV \subset W$. Similarly, g_2 is continuous if, and only if, for each neighborhood W of x^{-1}, there exists a neighborhood U of x such that $U^{-1} \subset W$.

It is easy to see that the mappings g_1 and g_2 are continuous in all their variables together if, and only if, the mapping

$$g_3 : (x, y) \to xy^{-1}$$

of $G \times G$ onto G is continuous.

Theorem 1. *Let a be a fixed element of a semitopological group G. Then the mappings*

$$r_a : x \to xa$$

$$l_a : x \to ax$$

of G onto G are homeomorphisms of G.

PROOF. It is clear that r_a is a 1:1 and onto mapping. Let W be a neighborhood of xa. Since G is a semitopological group, there exists a neighborhood U of x such that $Ua \subset W$. This shows that r_a is continuous. Moreover, it is easy to see that the inverse r_a^{-1} of r_a is the mapping: $x \to xa^{-1}$, which is continuous by the same argument as above. Hence, r_a is a homeomorphism. The fact that l_a is a homeomorphism follows similarly.

r_a and l_a are, respectively, called the *right* and *left* translations of G.

Corollary 1. *Let F be a closed, P an open, and A any subset of a semi-topological group G and let $a \in G$. Then:*
(i) *Fa, aF are closed.*
(ii) *Pa, aP, AP, and PA are open.*

PROOF. Since the mappings in Theorem 1 are homeomorphisms, (i) is obvious. By the same argument, Pa and aP are open in (ii). Since $AP = \bigcup_{a \in A} aP, PA = \bigcup_{a \in A} Pa$, and the union of open sets is open, (ii) is established.

Corollary 2. *Let G be a semitopological group. For any $x_1, x_2 \in G$, there exists a homeomorphism f of G such that $f(x_1) = x_2$.*

PROOF. Let $x_1^{-1}x_2 = a \in G$, and consider the mapping $f: x \to xa$. Then f is a homeomorphism by Theorem 1 and $f(x_1) = x_2$.
A space for which Corollary 2 is true is called a *homogeneous* space.

13. NEIGHBORHOOD SYSTEMS OF IDENTITY OF A SEMITOPOLOGICAL GROUP

In view of Theorem 1, §12, it follows that if one knows a fundamental system of neighborhoods of the identity of a semitopological group, then one can find a fundamental system of neighborhoods of any other point by translation. In the sequel, we show that actually the topology of a semi-topological group is completely determined by a fundamental system of neighborhoods of its identity. More precisely, we have the following:

Theorem 2. *If $\{U\}$ is a fundamental system of open neighborhoods of e in a semitopological group G, then $\{xU\}$ and $\{Ux\}$, where x runs over G and U over $\{U\}$, form bases of the topology of G.*
Conversely, let a filter base $\{U\}$ be given so that each U contains e and for each U and each $x \in U$ there exist V and W in $\{U\}$ such that $xV \subset U$ and $Wx \subset U$. Then there exists a topology u on G so that G, endowed with u, is a semitopological group.

PROOF. Let $a \in G$ and let W be an open neighborhood of a. Since $l_a^{-1}: x \to a^{-1}x$ is a homeomorphism in x (§12, Theorem 1), $l_a^{-1}(W) = a^{-1}W$ is an open set containing e and hence there exists a U in $\{U\}$ such that $U \subset a^{-1}W$. This implies $aU \subset W$, which proves that $\{xU\}$ is a base of the topology on G. Similar arguments show that $\{Ux\}$ is also a base.
Conversely, let $\tilde{\mathscr{U}}$ denote the family of all finite intersections of members in $\{U\}$. Then $\tilde{\mathscr{U}}$ is a nonempty family of U, each of which contains e.

Furthermore, for any $\tilde{U} = \bigcap_{i=1}^{n} U_i$, $x\tilde{U} = x\left(\bigcap_{i=1}^{n} U_i\right) = \bigcap_{i=1}^{n} xU_i$, for any $x \in G$.
And if $x \in \bigcap_{i=1}^{n} U_i$, then there exists a V_i for each i, $1 \leq i \leq n$, such that $xV_i \subset U_i$ and, hence, $\bigcap_{i=1}^{n} xV_i = x\left(\bigcap_{i=1}^{n} V_i\right) \subset \bigcap_{i=1}^{n} U_i$. This shows that the family $\tilde{\mathscr{U}}$ also satisfies the conditions assumed for the filter base $\{U\}$. By the definition of a subbase, the family of finite intersections of the family $\{xU\}$, where x runs over G and U over $\{U\}$, forms a base of the topology u on G. Now if $y \in \bigcap_{i=1}^{n} x_i U_i$, then $x_i^{-1}y \in U_i$ for each i, $1 \leq i \leq n$, and hence, by assumption, there exists a $V_i \in \{U\}$ such that $x_i^{-1}yV_i \subset U_i$, or $yV_i \subset x_i U_i$. This shows that

$$y\left(\bigcap_{i=1}^{n} V_i\right) = \bigcap_{i=1}^{n} yV_i \subset \bigcap_{i=1}^{n} x_i U_i.$$

Therefore $\{y\tilde{U}\}$, where $\tilde{U}$ runs over $\tilde{\mathscr{U}}$, forms a fundamental system of open neighborhoods of y for each $y \in G$. Similarly, one shows that $\{\tilde{U}y\}$ is also a fundamental system of neighborhoods of y.

Now to complete the proof, we have to show that G, endowed with u, is a semitopological group. Consider the mapping $g_1 : (x, y) \to xy$. Assume first that x is fixed, and let $\tilde{U}$ be any member in $\tilde{\mathscr{U}}$. Then $xy\tilde{U}$ is a member of a fundamental system of neighborhoods of xy. Since $y \in y\tilde{U}$ and $y\tilde{U}$ is a u-neighborhood of y as shown in the previous paragraph, $x(y\tilde{U}) \subset xy\tilde{U}$. This proves the continuity of g_1 in y while x is kept fixed. Similarly, one proves the continuity of g_1 in x by considering $\tilde{U}xy$ as a neighborhood of xy. This completes the proof.

A similar theorem, true for topological groups, will be proved in Chapter III.

14. CONSTRUCTIONS OF NEW SEMITOPOLOGICAL GROUPS FROM OLD

Many interesting results concerning the separation axioms, connectedness, etc., require continuity of group operations in both variables together as well as the continuity of inversion. Therefore, such results are postponed until the next chapter. However, a few results will show that one can define subgroups, quotient groups and product groups of semitopological groups as well.

More specifically, let G be a semitopological group and H a subgroup of G. Then H, endowed with the topology induced from G, is called a *semitopological subgroup*. The fact that the group multiplication is separately continuous in H is easy to see.

Moreover, if G is a semitopological group and if H is an invariant subgroup of G, then G/H, the collection of all distinct cosets $\{xH\}$, $x \in G$, forms a group called a *quotient group* of G. Let φ denote the *canonical* mapping. As usual, one defines a *quotient topology* on G/H as follows: A set $\dot{W}$ in G/H is open if, and only if, $\lambda^{-1}(\dot{W})$ is an open subset of G. It is easy to see that with this topology G/H is a semitopological group. We shall have details later, in Chapter III.

Furthermore, if $G_\alpha (\alpha \in A)$ is a family of semitopological groups, then the Cartesian direct product $G = \prod_{\alpha \in A} G_\alpha$ is the set of all functions $x: A \to G$ such that $x(\alpha) = x_\alpha$ for each $\alpha \in A$. G is also a group with coordinatewise multiplication or addition as the case may be. We can define the *product topology* on $\prod_{\alpha \in A} G_\alpha$ as the one having as its subbase the family $\{p_\alpha^{-1}(U_\alpha)\}$, where p_α is the αth *projection* mapping: $G \to G_\alpha$ and U_α runs over the open sets of G_α for each α.

Proposition 2. *If, for each $\alpha \in A$, G_α is a semitopological group, so is the direct product $G = \prod_{\alpha \in A} G_\alpha$, endowed with the product topology.*

PROOF. We have to show only that the mapping: $(x, y) \to xy$ of $G \times G$ onto G is continuous in each variable separately. Let W be a neighborhood of $xy \in G$. Then there exists a member U of the base of the product topology such that $xy \in U \subset W$. But then $U = \prod_{\alpha \in A} U_\alpha$, where $U_\alpha = G_\alpha$ for all $\alpha \in A$ except for a finite subset B of A, and for $\beta \in B$, U_β is an open set containing $x_\beta y_\beta = p_\beta(xy)$, $x_\alpha, y_\alpha \in G_\alpha$ for all α. Now, since each G_α is a semitopological group, for each $\beta \in B$ there exists a neighborhood V_β of y_β such that $x_\beta V_\beta \subset U_\beta$. But then $V = \prod_{\alpha \in A} V_\alpha$, where $V_\alpha = G_\alpha$ for $\alpha \in A \sim B$ and $V_\alpha = V_\beta$, $\alpha \in B$, is a neighborhood of y in G. Moreover, $p_\alpha(xy) = x_\alpha y_\alpha \in x_\alpha V_\alpha = x_\alpha p_\alpha(V) \subset U_\alpha$ for each $\alpha \in A$ shows that $xV \subset U \subset W$. This proves the continuity of $(x, y) \to xy$ when x is kept fixed. Similarly, one proves the continuity in y.

Proposition 3. *For each α, p_α is a continuous and open homomorphism of $G = \prod_{\alpha \in A} G_\alpha$ onto the semitopological group G_α.*

PROOF. For $x, y \in G$, $p_\alpha(xy) = x_\alpha y_\alpha = p_\alpha(x)p_\alpha(y)$. Therefore, p_α is a homomorphism. Since for each $x_\alpha \in G_\alpha$, $x = (x_\beta)$, where $x_\beta = e_\alpha$, $\beta \neq \alpha$, $x_\beta = x_\alpha$, $\beta = \alpha$, is an element of G and $p_\alpha(x) = x_\alpha$, it shows that p_α is onto. As remarked (§9), each projection mapping p_α is continuous and open.

Theorem 3. *If $G_\alpha (\alpha \in A)$ is a compact semitopological group for each α, so is $G = \prod_{\alpha \in A} G_\alpha$.*

PROOF. By Proposition 2, G is a semitopological group. It is compact owing to Tychonoff's theorem.

15. EMBEDDINGS OF ANY GROUP IN A PRODUCT GROUP

If for each α ($\alpha \in A$, any index set), $G_\alpha = G$, the product $\prod_{\alpha \in A} G_\alpha$ is denoted simply by G^A. Clearly, G^A is the set of all mappings of A into G. If G is a group so is G^A, as pointed out in §14.

Now if $A = G$, which is a group, then G^G is also a group.

Proposition 4. *Any group G can be embedded into G^G, i.e., there exists a one-to-one mapping of G onto a subset of G^G.*

PROOF. For each $a \in G$, let r_a denote the right translation of G onto G, i.e., $r_a : x \to xa$. Clearly, $r_a \in G^G$. Define $\eta_r : a \to r_a$, then it is easy to see that η_r is a $1:1$ mapping. Thus, G is mapped onto $\eta_r(G) \subset G^G$. In other words, G can be identified with the set of all its right translations. Similarly, one can identify G with the set of all its left translations as well.

Notations. The mappings $\eta_r : a \to r_a$ and $\eta_l : a \to l_a$ of G into G^G will be called the *right* and *left canonical embeddings* of G into G^G, respectively. In case G is an abelian group, $r_a = l_a$ and thus $\eta_r = \eta_l$.

Theorem 4. *Let G be a semitopological group. Then G_u is homeomorphic with $\eta_r(G) \subset G^G$, when the latter is endowed with the product topology.*

PROOF. We have to show that η_r is continuous and open. Let $T(\{x\}, U)$ be a member of the subbase of the product topology such that $r_a \in T(\{x\}, U)$, where
$$T(\{x\}, U) = \{f \in G^G : f(x) \in U\}$$

for an open set U in G and $x \in G$. Clearly, $r_a \in T(\{x\}, U)$ if, and only if, $r_a(x) = xa \in U$ or $a \in x^{-1}U$. Since G_u is a semitopological group and since U is an open set in G, $x^{-1}U$ is also open in G. Therefore, it follows that

$$\eta_r(x^{-1}U) = T(\{x\}, U) \cap \eta_r(G).$$

From this one sees easily that η_r is a homeomorphism.

Remark. If we agree to identify G with its right canonical image $\eta_r(G)$ into G^G, then the above theorem says that the product topology of G^G induces the initially given topology on a semitopological group G.

In view of Theorem 1, §12, every r_a or l_a is continuous if G is a semitopological group. Now let $C(G, G)$ denote the set of all continuous mappings of G into G. Then $\eta_r(G) \subset C(G, G) \subset G^G$, and thus we actually have more than what is asserted in Theorem 4, viz.:

Theorem 5. *Let G_u be a semitopological group. Then G_u is homeomorphic with a subset of $C(G, G)$, when the latter is endowed with the relative topology induced from the product topology on G^G.*

16. $\mathfrak{S}$-TOPOLOGIES AND SEMITOPOLOGICAL GROUPS

Let u and v be two topologies on a set E. Let $\{V_x\}$ denote a fundamental system of v-neighborhoods of an arbitrary point $x \in E$. Let $Cl_u V_x$ denote the u-closure of V_x. Let $u(v)$ denote the topology on E which has $\{Cl_u V_x\}$ as a fundamental system of neighborhoods of x, where V_x runs over $\{V_x\}$ and x over E.

Since for each $Cl_u V_x$ in $\{Cl_u V_x\}$, $Cl_u V_x \supset V_x$, the topology $u(v)$ is coarser than v. In symbols, either $u(v) \subset v$ or $v \supset u(v)$.

Proposition 5. *If E_v is a regular topological space and u any other topology on E such that $u \supset v$, then $v = u(v)$.*

PROOF. Let $\{V_x\}$ be a fundamental system of v-neighborhoods of an arbitrary point $x \in E$. Since E_v is regular, it can be assumed that each V_x is v-closed. But then $u \supset v$ implies that each V_x is u-closed as well and therefore $Cl_u V_x = V_x$ for each V_x in $\{V_x\}$. This proves that $u(v) = v$.

Let G_u be a semitopological group and, as usual, η_r the right canonical embedding of G_u into $C_p(G, G)$. First we define an $\mathfrak{S}$-topology on $C(G, G)$ as follows: Let $\{U\}$ be the collection of open sets of the topology u on G. Let $\mathfrak{S}$ denote the family of all closed subsets M of G. The family $\{T(M, U)\}$, where

$$T(M, U) = \{f \in C(G, G) : f(M) \subset U\},$$

M runs over $\mathfrak{S}$ and U over $\{U\}$, forms a subbase of a topology henceforth denoted by c.

If G_u is a T_1-space, then each singleton is a closed set and thus the relative product topology induced from G^G on $C(G, G)$ is coarser than c, i.e., $p \subset c$, if p also denotes the relative topology on $C(G, G)$.

If G is identified with $\eta_r(G) \subset C(G, G)$, then it follows that c induces a finer topology than $u = p$ on G. Thus, $u = p \subset c$ on G.

Remark. If $\mathfrak{S}'$ denotes the family of all closed sets with nonempty interiors, and if c' denotes the $\mathfrak{S}'$-topology, then clearly $c' \subset c$. For regular semitopological groups, one can show that $c' \supset p = u$, and for normal semitopological groups $c = c' \supset p = u$ (Husain[23]).

Proposition 6. *If G_u is a semitopological group, so is G_c, where G_c denotes the point set G endowed with the relative topology c induced from $C(G, G)$ when G and $\eta_r(G)$ (or $\eta_l(G)$) are identified.*

PROOF. Consider the mapping $f:(x, y) \to xy$. Let y be kept fixed first and let $T(M, U)$ be a member of the subbase of the topology c so that $xy \in T(M, U)$. In view of the canonical embedding, we can regard xy as the right translation r_{xy}. Then $xy \in T(M, U)$ is equivalent to $r_{xy}(M) = Mxy \subset U$, and hence $Mx \subset Uy^{-1}$. Since G_u is a semitopological group and since U is u-open, Uy^{-1} is u-open. Hence, $r_x(M) = Mx \subset Uy^{-1}$ shows that $x = r_x \in T(M, Uy^{-1})$ which, being a member of the subbase of the topology c, is a c-open neighborhood of r_x. Clearly, for all $z = r_z \in T(M, Uy^{-1})$, $Mz \subset Uy^{-1}$ implies $Mzy \subset U$ which proves the continuity of f in x while y is kept fixed. In a similar fashion, by making use of η_l, we can show the continuity of f in y while x is kept fixed.

Proposition 7. *Let G_u be a semitopological group and let G_c denote the point-set G endowed with the relative topology c induced from $C(G, G)$ under the canonical embedding η_r or η_l. If G_u is a normal topological space then $u(c) = c$.*

PROOF. From the above discussion it follows that $c \supset p = u$ on G and, therefore, $c \supset u(c) = p(c)$. To show the reverse inequality, let $x \in G$ and P be a c-neighborhood of x. Then there exists a finite collection of M_i $(1 \leq i \leq n)$ in $\mathfrak{S}$ and U_i $(1 \leq i \leq n)$ in $\{U\}$ such that

$$r_x = x \in \bigcap_{i=1}^{n} T(M_i, U_i) \subset P,$$

because the family of finite intersections of $\{T(M, U)\}$ is a base of c. Hence we have

$$r_x(M_i) = M_i x \subset U_i \text{ for each } i, 1 \leq i \leq n.$$

Since G_u is normal and since $M_i x$ is closed (because M_i is) and G_u is a semitopological group, there exists a u-open set V_i for each i $(1 \leq i \leq n)$ such that

$$M_i x \subset V_i \subset Cl_u V_i \subset U.$$

This shows that

$$r_x = x \in T(M_i, V_i) \subset T(M_i, Cl_u V_i) \subset T(M_i, U_i),$$

for each i, $1 \leq i \leq n$. Hence we have

$$r_x \in \bigcap_{i=1}^{n} T(M_i, V_i) \subset \bigcap_{i=1}^{n} T(M_i, Cl_u V_i) \subset \bigcap_{i=1}^{n} T(M_i, U_i) \subset P.$$

Since $Cl_u V_i$ is u-closed, $T(M_i, Cl_u V_i)$ is p-closed in G^G and so is $\bigcap_{i=1}^{n} T(M_i, Cl_u V_i)$. But then

$$x \in Cl_p \bigcap_{i=1}^{n} T(M_i, U_i) \subset \bigcap_{i=1}^{n} T(M_i, Cl_u V_i) \subset P.$$

Since $\bigcap_{i=1}^{n} T(M_i, V_i)$ is a member of the base of the topology c and, hence, $Cl_p \bigcap_{i=1}^{n} T(M_i, U_i)$ is a $p(c)$-neighborhood of x, it follows that $c \subset p(c) = u(c)$, because $u = p$ on G. Thus, combining the two inclusion relations between c and $p(c)$, we have $c = u(c)$ on G.

Remark. In view of the remark after Proposition 5, the above proposition implies that $c = c' = u(c') = u(c) = p(c)$ for normal semitopological groups.

Proposition 8. *Let G_u be a regular semitopological group and let G_c denote the set as indicated in Proposition 7. Then the mappings*
   (i) $f : (x, y) \to xy$ *of $G_u \times G_c$ onto G_u*
   (ii) $g : x \to x^{-1}$ *of G_c onto G_c*
are continuous.

PROOF. To show that f in (i) is continuous, let U be a u-open neighborhood of xy. Since for fixed y, $f(x, y) = r_y(x) = xy$ and r_y is continuous, there exists a closed u-neighborhood (owing to the regularity of G_u) M of x such that $r_y(M) = My \subset U$, which implies that $r_y \in T(M, U)$. Since $T(M, U)$ is a member of the subbase of the topology c, it is a c-neighborhood of r_y. Therefore, for $r_z = z \in T(M, U) \cap \eta_r(G)$, $r_z(M) = Mz \subset U$ shows that

$$M[T(M, U) \cap \eta_r(G)] \subset U.$$

This proves the continuity of f.
   For (ii), one sees that $r_x = x \in T(M, U)$ if, and only if, $r_x^{-1} \in T(G \sim U, G \sim M)$. For, r_x being $1:1$ and onto, $r_x(M) = Mx \subset U$ is equivalent to $r_x(G \sim M) = G \sim r_x(M) \supset G \sim U$ and hence equivalent to

$$r_x^{-1}(G \sim U) \subset G \sim M.$$

Since $T(M, U)$ and $T(G \sim U, G \sim M)$ are members of the subbase of the topology c, g is a continuous mapping of G_c onto G_c.

Theorem 6. *Let G_u be a regular semitopological group. If the topology c induced from $C(G, G)$ coincides with u on G, then G is a topological group.*

PROOF. If $u = c$ on G, then in Proposition 8, mappings (i) and (ii) are continuous when c is replaced by u. But this means precisely that G_u is a topological group.

17. B- AND C-TYPES OF SEMITOPOLOGICAL GROUPS

Definition 2. (a) A semitopological group G_u is said to be of *B-type* if for the topology c on G induced from $C(G, G) \subset G^G$, $c \supset u$ and $u(c) = c$ imply $c = u$.

(b) A semitopological group G_u is said to be of *C-type* if for each semi-topological group E_w and a $1:1$ continuous homomorphism f of E_w onto G_u, there exists a residual subset H of E such that the restriction $f \mid H: H_w \to G_c$ is continuous, where the topology c is, as usual, induced from $C(G, G)$.

Remark. The word "B-type" is suggested by the fact that similar relations such as $c \supset u$ and $u(c) = c$ used in the definition also occur in a characterization of barreled as well as B-complete spaces (Husain[22]). The word "C-type" is suggested by the words "Categorically related" used by M. K. Fort[12] in another context. For this see the following sections.

Theorem 7. *Every normal semitopological group G_u of B-type is a topological group.*

PROOF. In view of Theorem 6, §16, it will be sufficient to show that $u = c$ on G. Proposition 7, §16 implies $u(c) = c$ because G_u is normal by hypothesis. Since $c \supset u$ on G in general, by the definition of B-type, $u = c$. This proves the theorem.

Theorem 8. *Every normal Baire semitopological group G_u of C-type is a topological group.*

PROOF. In view of Theorem 7, it will be sufficient to show that G_u is of B-type. By Proposition 6, §16, G_c is a semitopological group. Consider the identity mapping $i: G_u \to G_u$ which is continuous. Since G_u is of C-type, there exists a residual subset H of G such that the restriction $i \mid H: H_u \to G_u$ is continuous. We know $c \supset u$ on G. Since G_u is normal, $u(c) = c$ by Proposition 7, §16. Furthermore, since G_u is a Baire space, H is everywhere dense in G_u. Also since the identity mapping $H_u \to G_c$ is continuous, it follows that on H, the induced topology u is finer than the induced topology c. But then, in general, $c \supset u$ on G implies $u = c$ on H.

Now to show that $u = c$ on G, let $x \in G$. Let $Cl_u V$ be a member of the topology $u(c)$ that contains x, where V is a member of a fundamental system of c-neighborhoods of x. Then $Cl_u V \cap H$ is c-open and hence u-open in H. But H being dense in G_u, $Cl_u V \supset Cl_u(Cl_u V \cap U)$, which is a u-neighborhood of x in G. This proves that $u(c) \subset u$. Since $c \supset u$, $u(c) = u$, and hence $u = c$ because $u(c) = c$.

We intend to derive some well-known particular cases of Theorem 8 and we shall require the following:

Lemma 1.   *Let u and v be two topologies on a set F such that:*
(a) u has a countable base $\{U_n\}$.
(b) For each $n \geq 1$, there exists a subset K_n in F such that $U_n \subset K_n$, and for any u-open set U, $x \in U$, there exists an n such that $x \in U_n \subset K_n \subset U$.
(c) If $x \in U_n$ for some n, there exists a v-open set V such that $x \in V$ and $V \sim K_n$ is v-open.
Then for each topological space E_w and a continuous mapping $f: E_w \to F_v$, there exists a residual (Chapter I, §2) subset H of E such that the restriction: $f \mid H: H_w \to F_u$ is continuous.

PROOF.   Define $D_n = \{x \in E:$ in each neighborhood of x there exist $y, z \in E$ such that $f(y) \in U_n$ and $f(z) \notin K_n\}$.
First of all, we show that all the points of E at which $f: E_w \to F_u$ is discontinuous are in $\bigcup_{n=1}^{\infty} D_n$. Let $x \in E$ be such a point. Then for some neighborhood U of $f(x)$, there exists no neighborhood W of x such that $f(W) \subset U$. Since $f(x) \in U$, (b) implies there exists an n such that $f(x) \in U_n \subset K_n \subset U$. Thus, it follows that each neighborhood of x contains points y and z such that $f(y) \in U_n$ and $f(z) \notin K_n$. In other words, $x \in D_n$.

For each n, D_n is a closed subset of E. For, let $x \in \overline{D_n}$. Then each open neighborhood of x contains elements of D_n. Hence, each neighborhood of x contains y and z such that $f(y) \in U_n$ and $f(z) \notin K_n$. In other words, $x \in D_n$.

Now we show that D_n, for each $n \geq 1$, is a nowhere-dense (Chapter I, §2) subset of E. Let us assume that D_n is not nowhere-dense for some n. Since D_n is closed, there exists a nonempty open set P such that $P \subset D_n$. By the definition of D_n, it follows that there exists $y \in P$ such that $f(y) \in U_n$. By (c), there exists a v-open set V such that $f(y) \in V$ and $V \sim K_n$ is v-open. Since $f: E_w \to F_v$ is continuous, there exists an open set W such that $y \in W \subset P$ and $f(W) \subset V$. But then there exists $z \in W$ (owing to the definition of D_n) such that $f(z) \notin K_n$ and, hence, $f(z) \in V \sim K_n$ because $f(z) \in V$. Again owing to continuity of $f: E_w \to F_v$, there exists an open set W_1 such that $z \in W_1 \subset W$ and $f(W_1) \subset V \sim K_n$, since the latter is a v-open neighborhood of $f(z)$. But by the definition of D_n, there exists a $t \in W_1$ such that $f(t) \in U_n$, which is impossible in view of the fact that $f(W_1) \subset V \sim K_n$ and $U_n \subset K_n$.

Hence, D_n is a nowhere-dense subset of E for each n, and therefore $\bigcup_{n=1}^{\infty} D_n$ is of the first category and contains all the points of u-discontinuity of f. In other words, at each point of $H = E \sim \bigcup_{n=1}^{\infty} D_n$, f is u-continuous. Clearly, H being a residual set, the lemma is established.
The above result is due to M. K. Fort.[12]

Theorem 9.   *A regular semitopological group G_u, satisfying the second axiom of countability, is of C-type.*

PROOF. Observe that every regular space satisfying the second axiom of countability is normal (Proposition 2, §5, Chapter I). Since G_u satisfies the second axiom of countability, u has a countable base $\{V_n\}$. Thus, the collection $\mathfrak{S}'$ of all closed subsets M in G such that either M is the complement of some V_n in $\{V_n\}$ or M has nonempty interior, consists of a countable family of sets. Therefore, the topology c' on $C(G_u, G_u)$ having the following sets:

$$U_n = \bigcap_{i=1}^{n} T(M_i, V_i)$$

as its base, has a countable base $\{U_n\}$. Thus, the family $\{U_n\}$ satisfies condition (a) of Lemma 1. Define $K_n = \bigcap_{i=1}^{n} T(M_i, \overline{V_i})$. Clearly, $U_n \subset K_n$ for each n. Furthermore, $p(c') = u(c') = c'$ on G (Proposition 7 and Remark, §16) because G_u is normal. From this it follows that condition (b) of Lemma 1 is satisfied. Since each K_n is p-closed, condition (c) is also satisfied. Hence, by Lemma 1, for each topological space E_w and a continuous mapping $f: E_w \to G_u$ there exists a residual subset H of E_w such that the restriction $f \mid H: H_w \to G_{c'}$ is continuous. Now taking $E_w = G_u$ and f as the identity mapping we see that the restriction of the identity mapping $i \mid H: H_u \to G_{c'}$ is continuous. But $c = c'$ on G, because G_u is normal (Remark after Proposition 5). Hence, $i \mid H: H_u \to G_c$ is continuous. This shows that G_u is of C-type, and the proof is complete.

Corollary 3. (Wu.[48]) *Every regular Baire semitopological group G_u satisfying the second axiom of countability is of B-type and hence a topological group.*

PROOF. This follows by combining Theorems 8 and 9, since a regular space satisfying the second axiom of countability is normal (Proposition 2, §5, Chapter I).

Corollary 4. (*Montgomery.*[33]) *A complete metric and separable semitopological group is a topological group.*

PROOF. This is immediate from Theorem 9, since each complete metric space is of the second category (Theorem 1, §2, Chapter I) and since each metric separable space satisfies the second axiom of countability (Chapter I, §4).

Corollary 5. *A regular locally compact semitopological group satisfying the second axiom of countability is a topological group.*

PROOF. This also follows from Theorem 9, since a Hausdorff locally compact space is a Baire space.

Corollary 6. *A metric compact semitopological group is a topological group.*

PROOF. This is a particular case of Corollary 4, since each metric compact space is separable and complete.

Corollary 7. *A Hausdorff compact semitopological group satisfying the second axiom of countability is a topological group.*

PROOF. This is a particular case of Corollary 5.

18. LOCALLY COMPACT SEMITOPOLOGICAL GROUPS

In this section we show that a Hausdorff locally compact semitopological group is a topological group if the mapping g_1 (Definition 1(a), §12) is continuous in both variables together. Thus, one sees that from Corollary 4, §17, the countability condition has been dropped under a certain condition. The results of this section are due to Ellis.[11]

We shall assume throughout this section that G_u is a Hausdorff locally compact semitopological group and that, in addition, the mapping: $(x, y) \to xy$ is continuous in both variables together.

Lemma 2. *Let A be a compact subset of G_u. Then A^{-1} is closed.*

PROOF. Let $x \in \overline{A^{-1}}$ and let $\{x_\alpha\}(\alpha \in N)$ be a net in A^{-1} such that $x_\alpha \to x$. Since $x_\alpha \in A^{-1}$, $x_\alpha^{-1} \in A$. Since A is compact, there exists a subnet $\{x_\beta^{-1}\}(\beta \in M \subset N)$ converging to $y \in A$. But then the continuity of the mapping g_1 (§12, Definition 1) in both variables together implies that $y = x^{-1}$ and, hence, $x \in A^{-1}$.

Lemma 3. *If x is a limit point of a countable subset E of G_u, then x^{-1} is a limit point of E^{-1}.*

PROOF. Let $E = \{x_1, x_2, \ldots, x_n, \ldots\}$. Then there is an ultrafilter $\mathscr{F}$ on E that converges to x. If we show that $\mathscr{F}^{-1}$ converges to y, then by continuity of the mapping g_1 it will follow that $y = x^{-1}$, and the proof will be complete.

Consider $F = E \bigcup \{x\}$ and $D = \bigcup\limits_{n=-\infty}^{\infty} F^n$. Then D is a countable subgroup of G. Let $A = \bar{D}$. Then by the continuity of multiplication, $A^2 \subset A$. Let C be a compact neighborhood of the identity. Then $A \subset DC^{-1}$, because $\bar{D} = A$. Hence, $A = \bigcup\limits_{d \in D} (dC^{-1} \cap A) = \bigcup\limits_{d \in D} d(C^{-1} \cap A)$, because D is a group and $A^2 \subset A$. But $\bar{D} = A$, being a closed subset of a locally compact space, is locally compact. Therefore, $d(C^{-1} \cap A)$ is closed by the application of Lemma 2, for each d. Now since D is countable and since A, being locally compact, is a Baire space, there exists an element d_0 for which $d_0(C^{-1} \cap A)$ has a nonempty interior. Therefore, there exists an open set $P \subset G$ such that $P \cap A \neq \varnothing$ and $P \cap A \subset d_0(C^{-1} \cap A)$. Since $\bar{D} = A$, there exists $c \in D \cap P$. Therefore, $xc^{-1}(P \cap A) = xc^{-1}P \cap A$ is a neighborhood of x relative to A. Since $\mathscr{F} \to x$ and $\mathscr{F}$ is an ultrafilter on A, there exists $K \in \mathscr{F}$ such that $K \subset xc^{-1}(P \cap A) \subset xc^{-1}d_0C^{-1}$ which implies that $K^{-1} \subset Cd_0^{-1}cx^{-1}$. But the latter is compact because C is compact. Therefore, $\mathscr{F}^{-1}$ converges to y.

Now we show that Lemma 2 can be improved, viz.:

Lemma 4. *If A is a compact subset of G_u, so is A^{-1}.*

PROOF. In view of Lemma 2, it will suffice to show that A^{-1} can be covered by a finite number of translates of a compact neighborhood V of the identity. Suppose this is not possible. Then there exists a sequence $\{x_n^{-1}\}$ in A^{-1} such that $x_n^{-1} \notin \bigcup_{i=1}^{n-1} x_i^{-1}V$. Let $E_n = \{x_k \in \{x_i\} : k \geq n\}$. By compactness of A, there exists x such that $x \in \bigcap_{n=1}^{\infty} \bar{E}_n$. Let U be a neighborhood of e such that $U^2 \subset V$. Then, since $x \in \bar{E}_1$, there exists $x_m \in E_1$ such that $x_m \in Ux$ and, hence, $x^{-1} \in x_m^{-1}U$. But $x \in \bar{E}_{m+1}$ implies that $x^{-1} \in \overline{E_{m+1}^{-1}}$ by Lemma 3. Thus, there is $n > m$ such that $x_n^{-1} \in x^{-1}U \subset x_m^{-1}U^2 \subset x_m^{-1}V$ which is a contradiction owing to the choice of $\{x_n^{-1}\}$.

Theorem 10. *A locally compact Hausdorff topological space that is also a group and in which the multiplication is a continuous function of two variables together is a topological group.*

PROOF. We have to show only that the mapping g_2 (§12, Definition 1(b)) is continuous. Let U be an open neighborhood of e. We wish to show that there exists a compact neighborhood C of e such that $C^{-1} \subset U$. Assume that for each compact neighborhood C of e, $C^{-1} \cap (G \sim U) \neq \varnothing$. Since C is compact so is C^{-1} by Lemma 4. Hence, the family $C^{-1} \cap (G \sim U)$, where C runs over all compact neighborhoods of e, forms a family of compact sets satisfying the finite intersection property. Therefore, $\bigcap (C^{-1} \cap (G_u \sim U)) \neq \varnothing$. But $\{e\} = \bigcap C^{-1}$, where C runs over all compact neighborhoods of e, and so

$$\bigcap (C^{-1} \cap (G \sim U)) \subset \bigcap C^{-1} = \{e\}.$$

Hence, $\{e\} = \bigcap (C^{-1} \cap (G \sim U))$ and so, in particular, $e \in G \sim U$ which is impossible because $e \in U$. This completes the proof.

An immediate consequence of the above theorem is the following:

Corollary 8. *Let G_u be a compact Hausdorff topological space that is also a group and in which the multiplication is continuous in both variables together. Then G_u is a topological group.*

Exercises

A. Semigroups

A set S is said to be a *semigroup* if for each $x, y \in S$, the composite $xy \in S$ and the composition is associative.

A semigroup that is also a topological space in which the composition is continuous in each variable separately, is called a *semitopological semigroup*.

A semigroup that is also a topological space in which the composition is continuous in both variables together is called a *topological semigroup*.

1. Show that a topological semigroup S is not necessarily a homogeneous space. (Hint: $S = [0, 1]$ be endowed with the topology induced from the reals.)

In a semigroup S, the *left* (or *right*) *cancellation law* holds if $zx = zy$ (or $xz = yz$) always implies $x = y$ for $x, y, z \in S$.

2. A compact topological semigroup satisfying the left and right cancellation laws is a topological group.

A topological semigroup S is said to be *monothetic* if there exists an $x \in S$ such that the set of all positive integral powers of x is dense in S. Such an x is called a *monothetic generator* of S (Hewitt and Ross[19]).

3. A compact monothetic topological semigroup with two distinct monoethic generators is a topological group.

B. Locally compact semitopological groups

1. Show that a compact Hausdorff semitopological group G_u is a topological group. (Hint: Show that the topology c on $C(G, G)$ induces the initial topology u on G when G_u is embedded into $C(G, G)$ by the canonical embedding mapping η_r or η_l (§15, Theorem 5), and use Theorem 6, §16. Also observe that the topology c, in this case, is the compact-open topology.) Also see Arens.[2]

2. Show that a locally compact Hausdorff semitopological group G_u is a topological group.

C. Semitopological linear spaces

Let E_u be a linear space on a real or complex field K and also, as a point-set, a topological space. E_u is said to be a *semitopological linear space* if the mappings:

(i) $(x, y) \rightarrow x + y$ of $E_u \times E_u$ onto E_u

(ii) $(\lambda, x) \rightarrow \lambda x$ of $K \times E_u$ onto E_u

are continuous in each variable separately, where $x, y \in E$ and λ is a scalar in the field K.

From (ii) follows that the mapping: $x \rightarrow -x$ of E_u onto E_u is continuous. A subset $M \subset E$ is said to be *circled* if $\lambda M \subset M$ for all $|\lambda| \leq 1$, $\lambda \in K$.

If (i) and (ii) are both continuous in all variables together, then E_u is said to be a *topological linear space*.

1. Show that a normal B-type semitopological linear space E_u is a topological linear space, if there exists a fundamental system of circled neighborhoods of the origin 0. (Hint: Use Theorem 7, §17, and then Proposition 5 of Bourbaki,[5] Chapter 1, §1, No. 3.) See Husain.[23]

2. Derive from (1) that a metric B-type semitopological linear space E_u is a topological linear space, if there exists a fundamental system of circled neighborhoods of 0 in E_u.

3. Show that a regular Baire semitopological linear space satisfying the second axiom of countability and having a fundamental system of circled neighborhoods of 0, is a topological linear space. (Hint: Use Theorem 9, §17.)

III

General Theory
of
Topological Groups

In this chapter we shall discuss the general theorems concerning topological groups. Separation axioms in a topological group, subgroups, quotient groups, and direct products of topological groups are treated. Some special theorems concerning connected groups are also given.

19. TRANSLATIONS IN TOPOLOGICAL GROUPS AND SOME EXAMPLES

Theorem 1. *An algebraic group G, endowed with a topology u, is a topological group if, and only if, the mapping: $(x, y) \to xy^{-1}$ of $G_u \times G_u$ onto G_u is continuous in both variables together.*

PROOF. In view of the definition of a topological group (Chapter II, §12, Definition 1(b)), the theorem follows from the remark after Proposition 1, §12.

Theorem 2. *Let G_u be a topological group. Then the right and left translations r_a, l_a, the inversion mapping: $x \to x^{-1}$ and the inner automorphisms: $x \to axa^{-1}$ are all homeomorphisms.*

PROOF. The homeomorphism of r_a and l_a follows from Theorem 1, §12. For the inversion mapping, let $f(x) = x^{-1}$. Then clearly f is 1:1,

43

continuous and onto. Since $f^{-1}(x) = x^{-1}$ is continuous, f is open and hence f is a homeomorphism.

The inner automorphism as the composition of two homeomorphisms, $x \to xa^{-1}$ and $x \to ax$, is a homeomorphism.

As Corollary 1, §12, follows from Theorem 1, §12, so does the following from the above theorem:

Corollary 1.   *Let F be a closed, P an open, A any subset of a topological group G and a ∈ G. Then aF, Fa, F^{-1} are closed; aP, Pa, P^{-1}, AP, PA are all open.*

Thus, it follows that the products PQ and QP of two open sets P and Q are open. However, the product of two closed subsets in a topological group is *not* necessarily closed, as will be shown later on.

Since a topological group is a semitopological group, it is a homogeneous space by Corollary 2, §12.

Examples. Some of the most useful and well-known examples of topological groups are as follows:

1. Let $G = R$, the real line with addition as the group operation and the usual metric topology defined by $d(x, y) = |x - y|$. For each $\varepsilon > 0$, $|x| < \varepsilon/2$, $|y| < \varepsilon/2$ imply $|x + y| < \varepsilon$ and therefore addition is continuous. Similarly, one sees easily that the inversion: $x \to -x$ is continuous. Hence, R is an additive abelian topological group. It is a metric, complete, locally compact and noncompact topological group satisfying the second axiom of countability.

2. $G = R \sim \{0\}$, the set of all nonzero real numbers with multiplication as the group operation and with the topology induced from the set of real numbers. One verifies that G is a multiplicative abelian topological group.

3. $G = R^{+} = \{x \in R : x > 0\}$, with the topology induced from $R \sim \{0\}$. This is a subgroup of $R \sim \{0\}$.

4. $G =$ the quotient group of R by the subgroup Z of integers $= R/Z$ is the *circle group* with the quotient topology of the additive group of the reals. This group will be denoted by T, for *torus group* or, more precisely, one-dimensional torus group.

If one considers the mapping: $x \to e^{\pi i x}$, $i = \sqrt{-1}$, of the set of real numbers into the set of complex numbers of modulus unity, one sees that the torus group T is homeomorphic with the set of all complex numbers of unit modulus, or, in other words, with the unit circle of the complex plane. That accounts for the other name of T, viz., T is a circle group. T is a compact abelian topological group.

5. $G = R^{n}$, the n-dimensional Euclidean space with addition as the coordinate addition. G is an additive abelian group. The topology is the usual metric topology defined by the metric: $d(x, y) = \sqrt{\sum_{i=1}^{n} (x_i - y_i)^2}$, in

which $x = (x_1, x_2, \ldots, x_n)$ and $y = (y_1, y_2, \ldots, y_n)$. This is also a noncompact, locally compact topological group satisfying the second axiom of countability.

6. $G = R^n$ modulo the group of all $x = (x_1, \ldots, x_n)$, in which each x_i is an integer. The topology is the usual quotient topology of R^n. This is a compact abelian topological group. This is called the *n-dimensional torus group* and is denoted by T^n.

7. Let G be any algebraic group with the indiscrete topology (see §1, Chapter I). Since $GG^{-1} = G$, for any group G, the group operations are continuous and hence G, endowed with the indiscrete topology, is a topological group. An indiscrete space is not even a T_0-space and so there is a topological group which is *not* a T_0-space and, hence, *not* Hausdorff.

8. Let G be any algebraic group with the discrete topology (see §1, Chapter I). Since for any open set W containing xy^{-1}, $\{x\}$ and $\{y^{-1}\}$ are open neighborhoods of x and y^{-1}, respectively, it follows that G is a topological group.

In particular, the additive group of the integers Z with the topology induced from the real numbers is a discrete, metric, countable, locally compact topological group satisfying the second axiom of countability. Note that Z is a closed subgroup of R. On the other hand, the additive group R of the real numbers with the discrete topology is a discrete, metric, noncountable, locally compact topological group failing to satisfy the second axiom of countability.

9. Let X be any topological space and F a topological group. Let $G = C(X, F)$ denote the set of all continuous mappings of X into F. For each pair $f, g \in C(X, F)$ and $x \in X$, define $f + g(x) = f(x) + g(x)$ or $fg(x) = f(x)g(x)$ depending upon whether F is an additive or multiplicative group. Then G is a group. If we endow G with the relative product topology induced from F^X, then G is a topological group. We shall discuss this example in detail later on.

20. NEIGHBORHOOD SYSTEMS OF IDENTITY

Definition 1. A subset U of a group G is said to be *symmetric* if $U = U^{-1}$. In case G is an additive group, U is symmetric if $U = -U$.

Proposition 1. *In a topological group there exists a fundamental system $\{U\}$ of symmetric neighborhoods of e.*

PROOF. Let $\{V\}$ be a fundamental system of open neighborhoods of e. Since $e = e^{-1}$, Theorem 2, §19 shows that for each V in $\{V\}$, V^{-1} is an open neighborhood of e. But $U = V \cap V^{-1}$ is a symmetric neighborhood of e because $U^{-1} = V \cap V^{-1} = U$. Therefore, each V contains a U. On the other hand, each neighborhood of e contains a V and so $\{U\}$ is a fundamental system of symmetric neighborhoods of e.

The above Proposition is not necessarily true in semitopological groups, since the inversion mapping is not continuous.

Proposition 2. *For each neighborhood W of the identity e in a topological group, and for each finite set $\{\varepsilon_i\}(1 \leq i \leq n)$, $\varepsilon_i = \pm 1$, there exists a symmetric neighborhood U of e such that $U^{\varepsilon_1} U^{\varepsilon_2} \cdots U^{\varepsilon_n} \subset W$.*

PROOF. In view of the preceding proposition, the assertion follows from the repeated use of the continuity of the mapping: $(x, y) \to xy^{-1}$.

Proposition 3. *Let $\{U\}$ be the system of all neighborhoods of e in a topological group G. Then for any subset A of G,*

$$\bar{A} = \bigcap AU = \bigcap UA.$$

PROOF. For any $x \in \bar{A}$ and $U \in \{U\}$, xU^{-1} being a neighborhood of x, $xU^{-1} \cap A \neq \varnothing$ and, therefore, $x \in AU$. Hence, $\bar{A} \subset \bigcap AU$. Conversely, assume $x \in AU$ for each U in $\{U\}$. Then for any open neighborhood P of x, $P^{-1}x$ is a neighborhood of e; so therefore, $x \in AP^{-1}x$ because $P^{-1}x \in \{U\}$. This implies that $x = ap^{-1}x$ for some $a \in A$ and $p \in P$. But then it follows that $a = p$ or, in other words, $P \cap A \neq \varnothing$. Hence, $x \in \bar{A}$. This completes the proof.

The above proposition says that for any subset A, $\bar{A} \subset AU$ and $\bar{A} \subset UA$ for each neighborhood U of e.

Corollary 2. *Let U be any neighborhood of e in a topological group G. Then there is a neighborhood V of e such that $\bar{V} \subset U$. In other words, each topological group is a T_3-space (§5).*

PROOF. We choose V such that $VV \subset U$ (Proposition 2). By Proposition 3, $\bar{V} \subset VV \subset U$. By Theorem 2, §19, this is true at each $x \in G$ and, hence, G is a T_3-space.

Theorem 3. *In each topological group G, there exists a fundamental system $\{U\}$ of closed neighborhoods of the identity e such that:*
(a) *Each U is symmetric.*
(b) *For each U in $\{U\}$ there exists a V in $\{U\}$ such that $V^2 \subset U$.*
(c) *For each U in $\{U\}$ and $a \in G$, there exists a V in $\{U\}$ such that $V \subset a^{-1}Ua$ or $aVa^{-1} \subset U$.*
Conversely, given a group G with a filter base $\{U\}$ satisfying (a)–(c), then there exists a unique topology u on G such that G_u is a topological group and $\{U\}$ forms a fundamental system of neighborhoods of e.

PROOF. Since G is a topological group, in view of Proposition 1 and Corollary 2, there exists a fundamental system $\{U\}$ of neighborhoods of e

such that each U is symmetric and closed. Thus, condition (a) is satisfied by $\{U\}$. Furthermore, owing to Theorem 1, §19, (b) is satisfied by virtue of the definition of a topological group; and (c) follows because of Theorem 2, §19.

Conversely, suppose $\{U\}$ is a filter base satisfying conditions (a)–(c) stated in the theorem. For each U in $\{U\}$, by (a) and (b) there exists a V in $\{U\}$ such that $VV^{-1} \subset U$. Hence, if $x \in V$, then $e = xx^{-1} \in VV^{-1} \subset U$ shows that each U in $\{U\}$ contains e. Therefore, each member of the systems $\{xU\}$ and $\{Ux\}$ contains x for any $x \in G$. One sees easily (Theorem 2, §13) that $\{xU\}$ and $\{Ux\}$ form a filter base at x, since $\{U\}$ is a filter base at e. Thus by Proposition 1, §3, there exists on G a unique topology u having $\{xU\}$ and $\{Ux\}$ as fundamental systems of neighborhoods of x for each $x \in G$.

Now to show that the mappings g_1 and g_2 (Definition 1, §12, Chapter II) are continuous, let $a, b \in G$ and suppose $x = ar$, $y = bs$, $r, s \in G$. Consider

$$(ab^{-1})^{-1}(xy^{-1}) = brs^{-1}b^{-1}.$$

Let P be a neighborhood of e. Then there exists a U in $\{U\}$ such that $U \subset P$. Also $brs^{-1}b^{-1} \in U$ if $rs^{-1} \in b^{-1}Ub$. By (c) there exists a V in $\{U\}$ such that $V \subset b^{-1}Ub$. But by (a) and (b) there exists a W in $\{U\}$ such that $WW^{-1} \subset V \subset b^{-1}Ub$. Now if $r, s \in W$ then $rs^{-1} \in WW^{-1} \subset b^{-1}Ub$. Therefore,

$$(ab^{-1})^{-1}(xy^{-1}) \in U \subset P,$$

for all $r, s \in W$. This proves the continuity of the mapping: $(x, y) \to xy^{-1}$ and, hence, G_u is a topological group.

In case the group is additive, the conditions of Theorem 3 can be stated as follows:

(a′) Each U is symmetric, i.e., $U = -U$.

(b′) For each U in $\{U\}$, there exists a V in $\{U\}$ such that $V + V \subset U$.

(c′) For each U in $\{U\}$, there exists a V in $\{U\}$ such that $V \subset -a + U + a$ or $a + V - a \subset U$.

In case G is abelian, one can dispense with (c) and (c′).

Proposition 4. *Let F be a closed subset and C a compact subset of a topological group G such that $F \cap C = \varnothing$. Then there exists a neighborhood U of e such that:*

(i) $FU \cap CU = \varnothing$

(ii) $UF \cap UC = \varnothing$.

PROOF. To show (i), it is sufficient to show that there exists a neighborhood U of e such that $C \cap FUU^{-1} = \varnothing$. For each neighborhood U of e, let $F_U = \overline{FUU^{-1}}$. Then F_U being closed, by Proposition 3, §20, $F_U = \overline{F_U} = \bigcap FUU^{-1}V$, where V runs over the system of all neighborhoods of e. But then

$$F_U = \bigcap FUU^{-1}V = \bigcup FW = \bar{F} = F$$

because F is closed by hypothesis and $W = UU^{-1}V$ runs over the system of all neighborhoods of e as V does. Hence, by hypothesis,

$$F_U \cap C = F \cap C = \emptyset,$$

for each F_U. In other words, $\{G \sim F_U\}$ is an open covering of C when U runs over the system of neighborhoods of e. But then there is a finite subfamily F_{U_i} $(1 \le i \le n)$ of $\{F_U\}$ such that

$$\left(\bigcap_{i=1}^{n} F_{U_i} \right) \cap C = \emptyset,$$

because C is a compact subset. Let $W = \bigcap_{i=1}^{n} U_i$, which is a neighborhood of e. Clearly $WW^{-1} = \bigcap_{i=1}^{n} U_i W^{-1} \subset \bigcap_{i=1}^{n} U_i U_i^{-1}$ and, hence, $FWW^{-1} \subset \bigcap_{i=1}^{n} FU_i U_i^{-1}$. But then, by taking closures, we see that the latter inclusion implies

$$FWW^{-1} \subset F_W \subset \bigcap_{i=1}^{n} F_{U_i}.$$

Hence, it follows that $FWW^{-1} \cap C = \emptyset$ and, therefore, $FW \cap CW = \emptyset$. This establishes (i).

Similarly, one proves the existence of a V such that $VF \cap VC = \emptyset$. Now let $U = W \cap V$, then for this U, both (i) and (ii) are satisfied.

Corollary 3. *Let F be a closed subset and C a compact subset of a topological group G. Then FC and CF are closed.*

PROOF. Let $x \in G \sim FC$. Then $F^{-1}x \cap C = \emptyset$, and $F^{-1}x$ is closed (Corollary 1, §19) because F is so by hypothesis. Since C is compact, by Proposition 4, there exists a neighborhood U of e such that

$$F^{-1}xU \cap CU = \emptyset.$$

But this implies that $xUU^{-1} \cap FC = \emptyset$. Hence, xUU^{-1} being a neighborhood of x not meeting FC, it follows that FC is closed, since x is arbitrary. Similarly, CF is closed.

The above corollary is not true if C is a noncompact subset. For example, consider the additive abelian group R of real numbers with the usual metric topology (§19, Example 1). The group Z of integers is a closed subgroup of R (§19, Example 8). Similarly the set $Z\alpha$, where α is an irrational number, is also a closed subgroup of R. But $Z + Z\alpha$ is everywhere dense in R and hence nonclosed.

21. SEPARATION AXIOMS IN TOPOLOGICAL GROUPS

Theorem 4. *For a topological group G, the following statements are equivalent:*

(a) *G is a T_0-space*
(b) *G is a T_1-space*
(c) *G is a T_2-space or Hausdorff space*
(d) *$\bigcap U = \{e\}$, where U is a fundamental system of neighborhoods of e.*

PROOF. We shall show that (a) $\Rightarrow$ (b) $\Rightarrow$ (c) $\Rightarrow$ (d) $\Rightarrow$ (a).

For (a) $\Rightarrow$ (b), let $x \neq y$, x, $y \in G$. By (a) for at least one (say, x) of x and y, there exists an open neighborhood P of x such that $y \notin P$. Since $x^{-1}P = V$ is an open neighborhood of e, $V \cap V^{-1} = Q$ is an open symmetric neighborhood of e and therefore yQ is a neighborhood of y. Now $x \notin yQ$ because otherwise $x^{-1} \in Qy^{-1}$ and, hence, $x^{-1} \in Qy^{-1} \subset Vy^{-1} \subset x^{-1}Py^{-1}$. But this implies that $e = xx^{-1} \in xx^{-1}Py^{-1} = Py^{-1}$, or $y \in P$, which is a contradiction.

For (b) $\Rightarrow$ (c), let $x \neq y$, x, $y \in G$. By (b), $\{x\}$ is a closed set and therefore $P = G \sim \{x\}$ is an open neighborhood of y and hence $y^{-1}P$ is an open neighborhood of e. Let V be an open neighborhood of e such that $VV^{-1} \subset y^{-1}P$. Then yV is an open neighborhood of y. Let $Q = G \sim \overline{yV}$ which is an open set, and $x \in Q$. For otherwise $x \in \overline{yV}$ and, hence, $xV \cap yV \neq \varnothing$. But this shows that $x \in yVV^{-1} \subset y(y^{-1}P) = P$, which is a contradiction because $x \notin P$. Clearly $Q \cap yV = \varnothing$, $y \in yV$ and $x \in Q$, yV and Q are open sets. This proves (c).

For (c) $\Rightarrow$ (d), let $x \in U$ for each U in $\{U\}$ and assume $x \neq e$. Then (c) implies that there exists a neighborhood P of e such that $x \notin P$. But then there exists a U in $\{U\}$ such that $U \subset P$. We have the contradiction: $x \in U \subset P$ and $x \notin P$. Hence, $x = e$, and (d) is established.

For (d) $\Rightarrow$ (a), let $x \neq y$. Then $xy^{-1} \neq e$ and, hence, by (d) there exists a U in $\{U\}$ such that $xy^{-1} \notin U$. Thus, Uy being a neighborhood of y and $x \notin Uy$, (a) is proved.

Definition 2. A topological group G is said to be *metrizable* if there exists a countable fundamental system $U_n (n \geq 1)$ of neighborhoods of the identity e such that $\bigcap_{n=1}^{\infty} U_n = \{e\}$, and satisfying the conditions (a)–(c) of Theorem 3, §20.

For topological groups, metrizability usually means that there exists a left (or right) invariant metric d such that the given topology of the topological group is equivalent with the one induced by the metric. (A metric is said to be *left* or *right invariant* if $d(ax, ay) = d(x, y)$ or $d(xa, ya) = d(x, y)$ for all a, x, $y \in G$, respectively). Since we do not make use of the metric in the sequel, we shall not pause to establish the equivalence of Definition 2 and the usual concept of metrizability (Kakutani[26]). The existence of an invariant metric for metrizable topological groups G can be established by constructing a continuous function f as in Theorem 5 below, and then putting $d(x, y) = \sup_{a \in G} \{|f(xa) - f(ya)|\}$.

Although one can show that a Hausdorff topological group is completely regular by showing that it is a uniform space, we show complete regularity directly in the following:

Theorem 5. *A Hausdorff topological group G is completely regular and hence regular.*

Proof. Let $\{U\}$ be a fundamental system of neighborhoods of $e \in G$ and satisfying the conditions of Theorem 3, §20. Now let C be a closed subset of G such that $e \notin C$. Then $G \sim C = U_0$ (say) is an open neighborhood of e. For each integer $n > 0$ there exists a U_n in $\{U\}$ such that (i) $U_{n+1}^2 \subset U_n$ for all $n \geq 0$. Let D denote the set of all dyadic rational numbers in $[0, 1]$, i.e., for each $r \in D$ there exist integers $n, k \geq 0$ such that $r = k/2^n, k \leq 2^n$. For each $r \in D$, we define $V(r)$ by induction as follows: (ii) $V(1/2^n) = U_n$ for $n \geq 0$. Suppose $V(r)$ has been defined for all $r = \dfrac{k}{2^n}, k \leq 2^n$. Define (iii) $V(k'/2^{n+1}) = V(k/2^n)$ if $k' = 2k$ and (iv) $V(k'/2^{n+1}) = V(1/2^n)V(k/2^n)$ if $k' = 2k + 1$.

If $0 \leq k = 2m \leq 2^n$, we have

$$
\begin{aligned}
V(1/2^n)V(k/2^n) &= V(1/2^n)V(m/2^{n-1}) && \text{by (iii)} \\
&= U_n V(m/2^{n-1}) && \text{by (ii)} \\
&\subset U_n^2 V(m/2^{n-1}) && \text{(since } e \in U_n) \\
&\subset U_{n-1} V(m/2^{n-1}) && \text{by (i)} \\
&\subset V(1/2^{n-1})V(m/2^{n-1}) && \text{by (ii)} \\
&\subset V(k + 1/2^n) && \text{by (iv).}
\end{aligned}
$$

Therefore, (v) $V(1/2^n) V(k/2^n) \subset V(k + 1/2^n)$ for all $0 \leq k \leq 2^n, k = 2m$. Similarly one can prove (v) when $k = 2m + 1$. Thus (v) is true for all integers k such that $0 \leq k + 1 \leq 2^n$.

Now we show that for $r_1, r_2 \in D, r_1 < r_2$ implies $V(r_1) \subset V(r_2)$. Suppose $r_1 = k_1/2^{n_1}$ and $r_2 = k_2/2^{n_2}$. Then $k_1 2^{n_2} < k_2 2^{n_1}$ and hence $k_1 2^{n_2}/2^{n_1+n_2} < k_2 2^{n_1}/2^{n_1+n_2}$. Clearly if $m + 1 < 2^n$ then $V(m/2^n) \subset V(m + 1/2^n)$ by (v). But then it follows that

$$V(k_1 2^{n_2}/2^{n_1+n_2}) \subset V(k_1 2^{n_2} + 1/2^{n_1+n_2}) \subset \cdots \subset V(k_2 2^{n_1}/2^{n_1+n_2})$$

in p steps, where $k_1 2^{n_2} + p = k_2 2^{n_1}$. But since $r_1 = k_1 2^{n_2}/2^{n_1+n_2}$ and $r_2 = k_2 2^{n_1}/2^{n_1+n_2}$, we see $V(r_1) \subset V(r_2)$.

Now define a real-valued function f as follows:

$$
f(x) = \begin{cases} \inf \{r \in D : x \in V_r\} \\ 1 \text{ if } x \notin V_1. \end{cases}
$$

Since $e \in V_r$ for all $r \in D$ and since inf $D = 0$, we see $f(e) = 0$. Furthermore $V_1 = V(1/2^0) = U_0 = G \sim C$ and, hence, $f(C) = 1$. Clearly, by the definition of f, $0 \leq f(x) \leq 1$ for all $x \in G$. Thus, in order to complete the proof, we have just to prove that f is continuous.

Let $x \in G$ such that $f(x) = 1$. If $y \in V(1/2^n)x$ then $y \in G \sim V(k/2^n)$, $k < 2^n - 2$. For, otherwise $y \in V(k/2^n)$ and the symmetry of the V's implies $x \in V(1/2^n)y \subset V(1/2^n) V(k/2^n) \subset V(k + 1/2^n)$ by (v). Hence, $f(x) < 1$, which contradicts the assumption that $f(x) = 1$. Thus it follows that $1 - \dfrac{1}{2^{n-1}} = (2^n - 2)/2^n \leq f(y) \leq 1$. Hence, $|f(y) - f(x)| \leq 1/2^{n-1}$. If for a given $\varepsilon > 0$, a sufficiently large n is chosen such that $1/2^{n-1} < \varepsilon$, the continuity of f at x follows immediately. The continuity of f when $f(x) = 0$ is even simpler to prove.

Now suppose $0 < f(x) < 1$ for some $x \in G$. Then there exist integers $m, k, k < 2^m, m > n + 1$ such that $x \in V(k/2^m) \sim V(k - 1/2^m)$ because $f(x) = \inf \{r \in D : x \in V_r\}$ and D is dense in $[0, 1]$. Using (v) as in the previous paragraph, for each $y \in V(1/2^m)x$, $y \notin V(k - 2/2^m)$. But clearly $x \in V(k/2^m)$ implies $y \in V(k + 1/2^m)$ by (v). Hence, by the definition of f, $(k - 2)/2^m \leq f(y) \leq (k - 1)/2^m$. Since $(k - 1)/2^m \leq f(x) \leq k/2^m$, we have $|f(x) - f(y)| \leq 1/2^m < 1/2^n < \varepsilon$. Hence, by the same arguments as above, we have the continuity of f in all cases. Since the translations are homeomorphisms, the above construction can be carried out at any point $x \in G$ instead of e. This completes the proof of the theorem.

Although it is true that a Hausdorff topological group is completely regular, it is not necessarily normal, as shown by the following example (see Hewitt and Ross[19]).

Example. Let $G = Z^A$, where Z is the additive abelian group of all integers endowed with the usual topology and A is an *uncountable* set. Let G be endowed with the product topology. (We shall show later that the direct product of topological groups is a topological group.) Being the product of Hausdorff spaces, G is Hausdorff. To show that G is *not* normal we consider the following:

Let $P = \{x \in G : \text{for } n \neq 0 \text{ there is at most one } \alpha \in A \text{ for which } p_\alpha(x) = n\}$ ($p_\alpha : G \to Z_\alpha$ is the αth projection mapping); and $Q = \{x \in G : \text{for } n \neq 1$ there is at most one $\alpha \in A$ for which $p_\alpha(x) = n\}$.

Let $x_0 \notin P$. Then $p_{\alpha_1}(x) = p_{\alpha_2}(x) = n$, for some $n \neq 0$ and some $\alpha_1, \alpha_2 \in A$, $\alpha_1 \neq \alpha_2$. Hence, $W = \{x \in G : p_{\alpha_1}(x) = p_{\alpha_2}(x) = n\}$ is open in G and contains x_0 such that $P \cap W = \varnothing$. Therefore, P is closed. Similarly, Q is also closed.

Now let U and V be any two open sets in G such that $P \subset U$ and $Q \subset V$. In order to show that G is not normal, it will suffice to show that $U \cap V \neq \varnothing$. For this, we first define a sequence $\{x^{(n)}\}$ in G, a sequence $\{\alpha_k\}$ in A and an increasing sequence $\{m^n\}(n \geq 1)$ in Z such that $p_\alpha(x^{(n)}) = k$ if $\alpha = \alpha_k(1 \leq k \leq m_{n-1})$ and $p_\alpha(x^{(n)}) = 0$ otherwise. Also

$$x^{(n)} \in \{x \in G : p_{\alpha_k}(x) = k \quad \text{for } 1 \leq k \leq m_{n-1} \quad \text{and}$$

$$p_{\alpha_k}(x) = 0 \quad \text{for } m_{n-1} < k \leq m_n\} \subset U.$$

The construction is carried out as follows:

Let $x^{(1)} \in G$ such that $p_\alpha(x^{(1)}) = 0$ for all $\alpha \in A$. Then by definition of P, $x^{(1)} \in P \subset U$. Since U is open, there exist a finite number of α's, say, $\{\alpha_k\}(1 \leq k \leq m_1)$ such that

$$x^{(1)} \in \{x \in G : p_{\alpha_k}(x) = 0 \text{ for } 1 \leq k \leq m_1\} \subset U.$$

Now let $x^{(2)} \in G$ be defined as follows: $p_\alpha(x^{(2)}) = k$ if $\alpha = \alpha_k$ $(1 \leq k \leq m_1)$ and $p_\alpha(x^{(2)}) = 0$ otherwise. Then clearly $x^{(2)} \in P \subset U$, and, hence, there exist a finite number of α's, say, $\{\alpha_k\}(m_1 < k \leq m_2)$, distinct from those determined earlier and such that $x^{(2)} \in \{x \in G : p_{\alpha_k}(x) = k$ for $\alpha_k(1 \leq k \leq m_1)$, and $p_{\alpha_k}(x) = 0$ for $\alpha_k(m_1 < k \leq m_2)\} \subset U$. By continuing in this way, one finds the desired sequences by induction.

Now we define $y \in G$ as follows: Let $p_\alpha(y) = k$ if $\alpha = \alpha_k (k = 1, 2, \cdots)$ and $p_\alpha(y) = 1$ otherwise. Then $y \in Q \subset V$. Hence there exist a finite number of α's, say, $\alpha_1, \ldots, \alpha_s$ such that

$$\{x \in G : p_\alpha(x) = p_\alpha(y), \alpha = \beta_i (1 \leq i \leq s)\} \subset V.$$

Since $\{m_n\}(n \geq 1)$ is an increasing sequence of integers, there exists an n_0 such that $\alpha_k \notin \{\beta_i, 1 \leq i \leq s\}$ whenever $k > m_{n_0}$.

Finally, we define $z \in G$ as follows:

$$p_\alpha(z) = \begin{cases} k \text{ if } \alpha = \alpha_k, k \leq m_{n_0} \\ 0 \text{ if } \alpha = \alpha_k, m_{n_0} < k \leq m_{n_0+1} \\ 1 \text{ otherwise.} \end{cases}$$

Then $z \in \{x \in G : p_\alpha(x) = p_\alpha(y), \alpha = \beta_i (1 \leq i \leq s)\} \subset V$, and $z \in \{x \in G : p_\alpha(x) = k$ for $\alpha = \alpha_k(1 \leq k \leq m_{n_0})$ and $p_\alpha(x) = 0$ for $\alpha = \alpha_k(m_{n_0} < k \leq m_{n_0+1})\} \subset U$. This proves that $U \cap V \neq \varnothing$.

We shall show later (Chapter IV) that a locally compact Hausdorff topological group is normal.

22. UNIFORM STRUCTURE ON A TOPOLOGICAL GROUP

For a general theory of uniform spaces, the reader is referred to Kelley[27] or Weil.[47] The latter is the creator of the concept of uniform spaces.

Let G be a topological group with $\{U\}$ as the system of all neighborhoods of e. For each U in $\{U\}$, define

$$L(U) = \{(x, y) \in G \times G : x^{-1}y \in U \text{ or } y \in xU\}$$

and

$$R(U) = \{(x, y) \in G \times G : yx^{-1} \in U \text{ or } y \in Ux\}.$$

Clearly $(x, x) \in L(U)$ and $R(U)$, since U is a neighborhood of e. That shows that the diagonal $\{(x, x) : x \in G\} \subset L(U) \cap R(U)$ for each U in $\{U\}$.

For each neighborhood U of e, U^{-1} is also a neighborhood of e and therefore

$$L(U^{-1}) = \{(x, y) \in G \times G : x^{-1}y \in U^{-1}\}$$
$$= \{(y, x) \in G \times G : y^{-1}x \in U\} = L^{-1}(U).$$

For each U in $\{U\}$, there exists a V in $\{U\}$ such that $V^2 \subset U$. Hence, it follows easily that
$$L^2(V) = L(V)\,L(V) \subset L(U).$$

Indeed, for each pair U and V in $\{U\}$, $U \cap V$ is in $\{U\}$ and, therefore,

$$L(U \cap V) = L(U) \cap L(V).$$

Thus, the family $\{L(U)\}$ and, similarly, $\{R(U)\}$, where U runs over $\{U\}$, forms a base for a uniformity. Therefore, $\{L(U)\}$ and $\{R(U)\}$ define a *left* and *right* *uniform structure* on G, respectively. Summing up we have:

Proposition 5. *A topological group G_u is a uniform space and $\{L(U)\}$, $\{R(U)\}$, and $\{L(U) \cup R(U)\}$ induce the original topology u.*

It is clear that the diagonal is equal to $\cap L(U)$ if, and only if, $\cap U = \{e\}$. In other words, the uniform structure is Hausdorff if the topological group is Hausdorff and vice versa.

Since every Hausdorff uniform space is completely regular (e.g., see Köthe,[28] p. 50) and in view of Proposition 5, one can establish Theorem 5, §21, indirectly.

Since in any uniform space the concepts of completion and uniform continuity are defined, one sees that the same concepts can be defined in a topological group as well. Uniform continuity can be defined as follows:

Definition 3. Let G and H be two topological groups and f a mapping of G into H. Let $\{U\}$ and $\{V\}$ denote the open bases of neighborhoods of e (the identity of G) and e' (the identity of H), respectively. Then f is said to be *left* (or *right*) *uniformly continuous* if for each V in $\{V\}$, there exists a U in $\{U\}$ such that $[f(x)]^{-1}f(y) \in V$ (or $f(y)[f(x)]^{-1} \in V$) whenever $x^{-1}y \in U$ (or $yx^{-1} \in U$) for $x, y \in G$.

Clearly, the left (or right) translations on a topological group are left (or right) uniformly continuous.

Clearly, each convergent filter in a topological group is a Cauchy filter, as is true in any uniform space. But the converse is not true. Whenever the converse is true, a topological group is said to be *complete*.

Note that there are two uniform structures on a topological group in general. For abelian topological groups, of course, the left and right uniform structures coincide.

As for any Hausdorff uniform space, a noncomplete Hausdorff abelian topological group G can be completed to a complete Hausdorff topological group $\hat{G}$ such that G is dense in $\hat{G}$. $\hat{G}$ is called the *completion* of G. Note that the completion, with respect to one of its uniformities (right or left), of a noncomplete Hausdorff nonabelian topological group is not necessarily a topological group. One requires additional conditions (e.g., see Bourbaki,[4] Chapter 3).

23. SUBGROUPS

Let G be a topological group and H an algebraic subgroup of G . Let H be endowed with the relative topology induced from G. Then, since the mapping: $(x, y) \to xy^{-1}$ of $G \times G$ into G is continuous, so is its restriction from $H \times H$ into H. In other words, H endowed with the relative topology is a topological group. H is called a *topological subgroup* or simply a *subgroup* of G. Clearly, if G is a Hausdorff topological group so is H.

Since the mapping $(x, y) \to xy^{-1}$ of $G \times G$ onto G is continuous, by a criterion of continuity (§8, Chapter I), we see that for any subsets A, B of G, $\bar{A} \ \overline{B^{-1}} \subset \overline{AB^{-1}}$, and for any $a \in G$, $\overline{aAa^{-1}} = a\bar{A}a^{-1}$.

Proposition 6. *If H is a subgroup of a topological group G, so is $\bar{H}$. If H is an invariant subgroup of G, so is $\bar{H}$.*

PROOF. In view of the previous paragraph, the proposition follows immediately by recalling that a subset A of an algebraic group G is a subgroup if $AA^{-1} \subset A$, and by using the continuity of $(x, y) \to xy^{-1}$.

In particular, the subgroup $\overline{\{e\}}$ is an invariant subgroup of G. $\overline{\{e\}} = e$ if, and only if, G is a T_0-space (§21, Theorem 4).

Proposition 7. *Every open subgroup H of a semitopological group (hence of a topological group) G is closed.*

PROOF. For each $x \in G$, xH is open by Corollary 1, §12. Hence, $H = G \sim \bigcup xH$ is closed, because $\bigcup xH$ is open, where the union is taken over all pairwise disjoint cosets different from H.

Proposition 8. *A subgroup H of a topological group G is closed if, and only if, for some closed neighborhood U of the identity e, $H \cap U$ is closed in G.*

PROOF. If H is closed, then evidently $H \cap U$ is closed for any closed set U. For the converse, assume $H \cap U$ is closed in G for some closed neighborhood U of e. Choose a symmetric neighborhood V of e such that $V^2 \subset U$. Let $x \in \bar{H}$. Let $\{x_\alpha\}(\alpha \in D)$ be a net in H such that $\{x_\alpha\}$ converges to x. By Proposition 6, $x^{-1} \in \bar{H}$ and, hence, there is some $y \in Vx^{-1} \cap H \neq \emptyset$. There exists an α_0 such that for $\alpha \geq \alpha_0$, $x_\alpha \in xV$. Thus, for $\alpha \geq \alpha_0$, we have $yx_\alpha \in Vx^{-1}xV = V^2 \subset U$ and, hence, $yx_\alpha \in U \cap H$ because $y, x_\alpha \in H$ and H is a subgroup. Since $U \cap H$ is closed by hypothesis, $yx \in U \cap H$ because yx_α converges to yx. Hence, $yx \in H$. But then $x = y^{-1}yx \in H^2 = H$. This completes the proof.

Proposition 9. *The center of a Hausdorff topological group G is a closed invariant subgroup.*

PROOF. Let C denote the center. Then it is well-known (§11) that C is an invariant subgroup of G and so is $\bar{C}$ (Proposition 6). To show that C is closed, it will suffice to show that $\bar{C} \subset C$. Let $x \in \bar{C}$ and suppose there exists $a \in G$ such that $a^{-1}xa \neq x$. Since G is Hausdorff and hence regular (§21, Theorem 5), there exist open sets U and V such that $x \in U$, $a^{-1}xa \in V$ and $\bar{U} \cap \bar{V} = \varnothing$. Since $x \in \bar{C}$, it is easy to see that $x \in \overline{U \cap C}$. Hence, $a^{-1}xa \in \overline{a^{-1}U \cap Ca} = \overline{a^{-1}(U \cap C)a} = \overline{U \cap C} \subset \bar{U}$, because C is the center. But this is a contradiction, and the proof is complete.

Proposition 10.   *Let U be a symmetric neighborhood of e in a topological group G. Then $H = \bigcup\limits_{n \geq 1} U^n$ is an open and closed subgroup of G.*

PROOF.   Let $x, y \in H$. Then there exist positive integers n, m such that $x \in U^n$ and $y \in U^m$.   Hence, $xy \in U^n U^m = U^{n+m} \subset H$.   Furthermore, $x^{-1} \in (U^n)^{-1} = (U^{-1})^n = U^n$, because U is symmetric. Thus, H is a subgroup of G. Now to show that H is open, we observe that for each $y \in H$, $yU \subset yH = H$. This proves that H is open and closed by Proposition 7.

Theorem 6.   *For any neighborhood U of e in a connected topological group G, $G = \bigcup\limits_{n \geq 1} U^n$. In other words, each connected group is "generated" by a neighborhood of its identity.   But the converse is not true.*

PROOF.   In view of Proposition 1, §20, one can assume that U is symmetric.   By Proposition 10, $\bigcup\limits_{n \geq 1} U^n$ is an open and closed subgroup of G. Since G is connected, $G = \bigcup\limits_{n \geq 1} U^n$ because in a connected space the only closed and open sets are the total space and the null set.

For the converse, let G be the additive group Q of the rational numbers with the relative topology induced from the real numbers. Q is not connected. But every symmetric neighborhood U of 0 in Q, being the trace of a symmetric neighborhood of 0 in R that generates R, generates Q.

Proposition 11.   *The component of the identity of a topological group is a closed invariant subgroup.*

PROOF.   Let C be the component of e. Then C is closed in general. Now to show that it is an invariant subgroup, let $a \in C$. Then $a^{-1}C \subset C$ because $a^{-1}C$, being the image of the component C under the homeomorphism: $x \to a^{-1}x$, is connected and $e \in a^{-1}C$. Hence, $\bigcup\limits_{a \in C} a^{-1}C = C^{-1}C \subset C$. This shows that C is a subgroup. Furthermore, for $a \in G$, $a^{-1}Ca$ is also connected because the mapping: $x \to a^{-1}xa$ is continuous. Since $e \in a^{-1}Ca$ and the latter is connected, so $a^{-1}Ca \subset C$ for each $a \in G$ owing to the fact that C is the component. This completes the proof.

A subgroup H of a topological group G is said to be a *discrete subgroup* if for each $x \in H$, there exists a neighborhood U of x such that $U \cap H = \{x\}$.

Proposition 12. *Every discrete subgroup H of a Hausdorff topological group G is closed.*

PROOF. By Corollary 2, §20, there exists a closed neighborhood U of e in G such that $U \cap H = \{e\}$ which is closed in G, because G is Hausdorff. Hence by Proposition 8, H is closed.

Remark. This proposition shows that the group of the integers is a closed subgroup of the group of the real numbers.

Proposition 13. *Every invariant discrete subgroup H of a connected Hausdorff topological group G is a closed invariant subgroup of the center of G.*

PROOF. By Proposition 12, H is a closed subgroup of G. We have to show only that H lies in the center. Let $a \in H$. Then there exists a neighborhood U of a in G such that $U \cap H = \{a\}$. Let V be a symmetric neighborhood of e such that $y^{-1}ay \subset U$ for all $y \in V$. Since H is an invariant subgroup, $y^{-1}ay \in H \cap U = \{a\}$ shows that $ay = ya$ for all $y \in V$. Since G is connected, by Theorem 6, $G = \bigcup_{n \geq 1} V^n$. Therefore, for any $x \in G$, there exist y_i ($1 \leq i \leq n$), $y_i \in V$ such that $x = y_1 y_2 \cdots y_n$. But then

$$\begin{aligned} x^{-1}ax &= (y_1 \cdots y_n)^{-1}a(y_1 \cdots y_n) \\ &= y_n^{-1}y_{n-1}^{-1} \cdots y_1^{-1}ay_1 \cdots y_n \\ &= a \end{aligned}$$

because $y_i^{-1}ay_i = a$ for each y_i. This shows that a is in the center of G, and the proof is ended.

The following few results about subgroups are useful.

Theorem 7. *Let G be a Hausdorff topological group and H a subgroup of G. Then*
   (a) *H is Hausdorff.*
   (b) *H is a compact subgroup if G is compact and H is closed.*
   (c) *H is a locally compact subgroup if G is locally compact and H is closed.*
   (d) *H is metrizable if G is.*
   (e) *H satisfies the second axiom of countability if G does.*
   (f) *H is a complete subgroup if G is and H is closed.*

PROOF. (a) and (b) are generally known for any topological space. For (c), let U be a neighborhood of e in G such that $\bar{U}$ is compact. Then $U \cap H$ is a neighborhood of e in H. And the closure of $U \cap H$ taken in H is equal to the closure of $U \cap H$ taken in G, since H is closed. But $\bar{U}$ being compact, $\overline{U \cap U}$ is also compact because the latter is a closed subset of $\bar{U}$ This proves (c).

(d) and (e) are immediate from relativization, and (f) is well-known in any uniform space (Proposition 7, §10).

All parts from (a)–(e) of this theorem are true for semitopological groups as well.

24. QUOTIENT GROUPS

Let G be a semitopological group and H a subgroup of G. Let G/H denote the set of all cosets $\{xH\}$, $x \in G$. Let φ be the canonical mapping of G into G/H (Chapter I, §11). With the help of φ we have defined (Chapter II, §14) a topology on G/H as follows: A subset $\dot{A}$ of G/H is open if, and only if, $\varphi^{-1}(\dot{A})$ is an open subset of G. It is easy to check that the totality $\{\dot{A}\}$ of open sets thus defined defines a topology. This topology in G/H is called the *quotient topology* and G/H, endowed with the quotient topology, is called a *quotient space*.

Note that by definition $\dot{A} = \{aH : a \in A\}$ is open if, and only if, $\bigcup\limits_{a \in A} aH = AH$ is an open set in G.

The important facts about φ are collected in the following:

Theorem 8. *Let G be a semitopological group and H a subgroup of G. Let G/H be the quotient space, endowed with the quotient topology, and φ the canonical mapping of G into G/H. Then:*

(a) *φ is onto;*

(b) *φ is continuous;*

(c) *φ is open.*

(d) *The quotient topology is the finest topology on G/H with respect to which φ is continuous.*

PROOF. The truth of (a) is obvious and (b) follows by the definition of a quotient topology. For (c), let U be open in G. We must show that $\varphi(U)$ is open in G/H, i.e., $\varphi^{-1}(\varphi(U))$ is open in G. But $\varphi^{-1}(\varphi(U)) = \{x : x \in uH$ for some $u \in U\} = UH$, which is open (Corollary 1, §12).

For (d), let v be any other topology on G/H such that $\varphi : G \to G/H$ is continuous. Then for each v-open set V in G/H, $\varphi^{-1}(V) = VH$ is open in G. But by the definition of the quotient topology, all such V's are open in the quotient topology. This shows that the quotient topology is finer than v. Thus, the proof is completed.

Proposition 14. *Let G be a semitopological group and H a subgroup of G. Then G/H is a homogeneous space (see Remark after Corollary 2, §12).*

PROOF. Let $\dot{x}_1, \dot{x}_2 \in G/H$. Then $\dot{x}_1 = x_1 H$ and $\dot{x}_2 = x_2 H$. Let a be in G such that $ax_1 = x_2$. Define the mapping $f_a : \dot{x} = xH \to (ax)H$ for all $\dot{x} \in G/H$. Then f_a is well-defined and is a $1:1$ mapping of G/H onto itself as is easy to check. Also $f_a^{-1} : \dot{x} \to (a^{-1}x)H$. Now to show that f_a is open and continuous, it will suffice to show that f_a is open. Let $\dot{U}$ be an open set in G/H. Then $\dot{U} = \varphi(UH)$, where U is an open set in G. But $f_a(\dot{U}) = \varphi(aUH)$, which is open in G/H since φ is an open mapping and aUH is an open set in G. Hence, f_a is a homeomorphism. Clearly, $f_a(\dot{x}_1) = a\dot{x}_1 = (ax_1)H = x_2H = \dot{x}_2$ shows that G/H is a homogeneous space.

Proposition 15. *Let G be a semitopological group and H a subgroup of G. Then G/H is a T_1-space if, and only if, H is closed.*

PROOF. If G/H is a T_1-space then clearly each singleton of G/H is a closed subset and hence, in particular, H as a point in G/H is closed. Since φ is continuous, $H = \varphi^{-1}(H)$ as a set in G is closed. Conversely, if H is a closed subgroup, by Corollary 1, §12, xH is a closed set in G and, therefore, $G \sim xH$ is open in G. Thus, $\varphi(G \sim xH)$ is open in G/H and since $(G/H) \sim \{xH\} = \varphi(G \sim xH)$ we see that $\{xH\}$ is closed. Hence, G/H is a T_1-space.

Proposition 16. *Let G be a semitopological group and H a subgroup of G. Then G/H is a discrete topological space if, and only if, H is open.*

PROOF. If G/H is discrete then each subset, and hence each singleton, is open. In particular, H as a point in G/H is open and $H = \varphi^{-1}(H)$ as a set in G is open. On the other hand, if H is open, so is xH for each $x \in G$. This shows that each singleton $\{\dot{x}\}$ is an open set in G/H and, therefore, the latter is a discrete space.

Theorem 9. *Let H be a subgroup of a semitopological group G, and φ the canonical mapping of G onto G/H. If $\{U\}$ is a fundamental system of neighborhoods of e in G, then $\{\varphi(U)\}$ is a fundamental system of neighborhoods of $\dot{e} = \varphi(e)$ in G/H.*

PROOF. By Theorem 8, for each neighborhood U of e in G, $\varphi(U)$ is a neighborhood of $\dot{e}$. Now let $\dot{V}$ be any neighborhood of $\dot{e}$ in G/H. Then $\varphi^{-1}(\dot{V})$ is a neighborhood of $e \in G$ and so there exists a U in $\{U\}$ such that $U \subset \varphi^{-1}(\dot{V})$. Whence it follows that $\varphi(U) \subset \dot{V}$. This completes the proof.

Let G be a semitopological (or topological) group and H an *invariant* subgroup of G. Then from Chapter I, §11, it follows that G/H is a group and is called a *quotient group*. The element $H = eH = \dot{e}$ is the identity of G/H. For each $xH = \dot{x} \in G/H$, $(\dot{x})^{-1} = (xH)^{-1} = H^{-1}x^{-1} = Hx^{-1} = x^{-1}H = (\dot{x}^{-1})$, since H is an invariant subgroup. We show the following:

Theorem 10. *Let G be a semitopological (or topological) group and H an invariant subgroup of G. Then*
(i) *The canonical mapping $\varphi : G \to G/H$ is a continuous and open homomorphism.*

(ii) G/H, *endowed with the quotient topology, is a semitopological (or topological) group.*

(iii) *If G is a topological group, then G/H is a completely regular topological group if, and only if, H is a closed subgroup of G.*

PROOF. (i) By Theorem 8, φ is clearly a continuous and open mapping. We have to show only that φ is a homomorphism. For this let $x, y \in G$. Then

$$\varphi(xy) = xyH = xyH^2 = xHyH = \varphi(x)\varphi(y),$$

since H is an invariant subgroup of G.

(ii) When G is a topological group, we have to show that the mapping: $(\dot{x}, \dot{y}) \to \dot{x}\dot{y}^{-1}$ of $G/H \times G/H$ onto G/H, is continuous. Let $\dot{W}$ be an open neighborhood of $\dot{x}\dot{y}^{-1}$, where $\dot{x} = xH$, $\dot{y} = yH$, $x, y \in G$. Clearly, $\varphi^{-1}(\dot{W})$ is open in G and $xy^{-1} \in \varphi^{-1}(\dot{W})$. Since G is a topological group, there exist open sets U, V in G such that $x \in U$, $y^{-1} \in V^{-1}$ and

$$xy^{-1} \in UV^{-1} \subset \varphi^{-1}(\dot{W}).$$

Whence, in view of (i), we have

$$\dot{x}\dot{y}^{-1} \in \varphi(U)[\varphi(V)]^{-1} \subset \varphi(\varphi^{-1}(\dot{W})) = \dot{W}.$$

By (i), φ is open and $\varphi(U)$ and $[\varphi(V)]^{-1} = \varphi(V^{-1})$ are open because U and V^{-1} are open. This proves (ii). Similar arguments work for semitopological groups.

(iii) In view of Theorems 4 and 5, §21, and (ii), it is sufficient to show that G/H is a T_1-space if, and only if, H is closed. But this follows from Proposition 15.

Definition 4. If the component (§7, Chapter I) of the identity in a topological group consists of the identity only, then the group is called *totally disconnected.*

Proposition 17. *Let G be a topological (or semitopological) group and C the component of the identity. Then G/C is a totally disconnected Hausdorff (or T_1-) topological (or semitopological) group.*

PROOF. By Proposition 11, §23, C is a closed invariant subgroup. Hence, by Proposition 15 and Theorem 10, G/H is a T_1 topological or semitopological group. If G is a topological group, then G/H as a T_1-space is Hausdorff by Theorem 4, §21. We have to show only that G/C is totally disconnected. Let $\dot{U}$ be the component of the identity $\dot{e}$ in G/C. $\varphi^{-1}(\dot{U})$ is a subset of G and $\varphi^{-1}(\dot{U})$ contains C. If G/C is not totally disconnected, there exists $\dot{x} \neq \dot{e} = C$, $\dot{x} \in \dot{U}$. This means that C is a proper subset of $\varphi^{-1}(\dot{U})$. Since C is the maximal connected set containing e, $\varphi^{-1}(\dot{U})$ is not connected. Let P and Q be open sets in G such that

(1) $$\varphi^{-1}(\dot{U}) = (P \cap \varphi^{-1}(\dot{U})) \cup (Q \cap \varphi^{-1}(\dot{U}))$$

where $(P \cap \varphi^{-1}(\dot{U})) \cap (Q \cap \varphi^{-1}(\dot{U})) = \varnothing$, and neither set is empty. Hence

$$\dot{U} = (\varphi(P) \cap \dot{U}) \cup (\varphi(Q) \cap \dot{U}),$$

as is easy to check. Clearly, for an $x \in U$, (since $\varphi^{-1}(U) = UC$), $xC \subset UC$. Therefore, from (1) we have

$$xC = (P \cap xC) \cup (Q \cap xC).$$

Since xC is connected, either $xC \subset P \cap xC$ or $xC \subset Q \cap xC$. Consequently the images of $P \cap UC$ and $Q \cap UC$ under φ are disjoint, since they are unions of cosets of C. In other words, $(\varphi(P) \cap \dot{U}) \cap (\varphi(Q) \cap \dot{U}) = \varnothing$. Since φ is an open mapping, $\varphi(P)$ and $\varphi(Q)$ are open sets and thus we have shown that $\dot{U}$ is not connected. But this is contrary to the assumption that $\dot{U}$ is the component of $\dot{e}$. Therefore, G/C is totally disconnected.

A few useful cases in which a property of a topological group is preserved under quotient group formation are collected in the following:

Theorem 11. *Let G be a semitopological group and H an invariant closed subgroup of G. Then:*

    (a) *G/H is compact if G is;*
    (b) *G/H is locally compact if G is a locally compact topological group;*
    (c) *G/H satisfies the first axiom of countability if G does;*
    (d) *G/H satisfies the second axiom of countability if G does.*

PROOF. Clearly (a) is true since G/H is the image of a compact space G under the continuous mapping $\varphi: G \to G/H$.

(b) Let U be a neighborhood of e in G such that $\bar{U}$ is compact. By (i) of Theorem 10, $\varphi(U)$ is a neighborhood of $\dot{e}$, and $\varphi(\bar{U})$ is compact and therefore closed in G/H (H is a closed subgroup, whence G/H is Hausdorff). Since $\overline{\varphi(U)} \subset \overline{\varphi(\bar{U})} = \varphi(\bar{U})$, $\overline{\varphi(U)}$ as a closed subset of a compact set is compact. This proves (b).

(c) Let $\{U_n\}$ be a countable fundamental system of neighborhoods of e in G. Then by Theorem 9, $\{\varphi(U_n)\}$ forms a fundamental system of neighborhoods of $\dot{e}$ in G/H under the quotient topology. Since $\{\varphi(U_n)\}$ is countable, it follows that G/H satisfies the first axiom of countability at $\dot{e}$. But since the translations are homeomorphisms of G/H, G/H satisfies this axiom everywhere.

(d) If $\{B_n\}$ is a countable base of the topology of G, then by the same argument as used in Theorem 9, it follows that $\{\varphi(B_n)\}$ forms a base of the quotient topology in G/H.

In the remainder of this section we shall discuss the subgroups of quotient groups and the so-called first and second isomorphism theorems.

Proposition 18. *Let G be a topological group, H an invariant subgroup of G, M any subgroup of G, and $\varphi: G \to G/H$. Then $\varphi(M)$ is a topological subgroup of G/H, and it is homeomorphic with MH/H.*

PROOF. Clearly, MH is a subgroup of G, $MH \supset H$ and H is an invariant subgroup of G. Thus, H is an invariant subgroup of MH. We see MH/H is a group and, by an isomorphism theorem of abstract groups, $\varphi(M)$ is isomorphic to MH/H.

Now we must show that the isomorphism of $\varphi(M)$ and MH/H is a homeomorphism. For this, let $\dot{A} = \{\dot{x} \in MH/H : x \in A\}$ be open in MH/H. Then that means there exists an open set U in G such that $A = U \cap MH$. Hence, by an abuse of notation:

$$\dot{A} = \{\dot{x} : x \in U \cap MH\} = \{\dot{x} : x \in U\} \cap \{\dot{x} : x \in MH\}$$
$$= \varphi(UH) \cap \varphi(M).$$

Since UH is open in G, $\dot{A}$ is open in $\varphi(M)$. Conversely, suppose $\dot{B} = \{\dot{x} : x \in B\}$ is open in $\varphi(M)$. Then there exists an open set $\dot{U} = \varphi(U)$ in G/H such that $\dot{B} = \dot{U} \cap \varphi(M)$, where U is open in G. But $\dot{U} \cap \varphi(M) = \{\dot{x} : x \in U \cap MH\}$. Therefore,

$$\dot{B} = \varphi((U \cap MH)H) = \varphi(UH \cap MH^2) = \varphi(UH \cap MH).$$

Since UH is open in G, it follows that $\dot{B}$ is open in MH/H. This completes the proof.

As is known (Theorem 4, §11, Chapter I) from group theory, the groups MH/H and $M/(M \cap H)$ are isomorphic if H is an invariant subgroup of a group G and M is any subgroup of G. In general, however, these quotient groups are not topologically isomorphic. For example, let α be an irrational real number, $\alpha \in R$, $H = Z$ the additive group of integers, $M = \alpha Z$, then $M \cap H = \{0\}$ and $M + H$ is everywhere dense in R. It is easy to see that the compact group $(M + H)/H$ is not homeomorphic with the discrete group $M/\{0\}$. The following, however, is true:

Proposition 19. *Let H be an invariant subgroup of a topological group G and M any subgroup of G. Let $f(\dot{m}) = m(M \cap H)$, $m \in M$. Then f is an open mapping of MH/H onto $M/(M \cap H)$.*

PROOF. Let $\dot{A}$ be an open subset of MH/H. That means $\dot{A} = \varphi(AH)$, $A \subset M$, and AH is relatively open in $MH \subset G$. But $f(\dot{A}) = \{x(M \cap H) : x \in A\}$. Since $A(M \cap H) = AH \cap M$, $A(M \cap H)$ is relatively open in M. Therefore, by the definition of the quotient topology of $M/(M \cap H)$, $f(\dot{A}) = \{x(M \cap H) : x \in A\}$ is open in $M/(M \cap H)$.

The above proposition corresponds to the first law of isomorphism in the theory of abstract groups. The second law of isomorphism corresponds to the following:

Proposition 20. *Let G be a topological group, H and M two invariant subgroups of G such that $H \subset M$. Then G/M is homeomorphic with $(G/H)/(M/H)$.*

PROOF. Algebraically it is well-known (Chapter I, §11, Theorem 5) that G/M and $(G/H)/(M/H)$ are isomorphic. It is clear that M/H is an invariant subgroup of G/H. The canonical mappings $\varphi:G \to G/H$ and $\psi:G/H \to (G/H)/(M/H)$ are continuous, open and onto homomorphisms. Therefore, the composition mapping $\psi \circ \varphi:G \to (G/H)/(M/H)$ is also a continuous, open, and onto homomorphism. But it is easy to check that the kernel of $\psi \circ \varphi$ is M. Hence, by Proposition 3, §30, Chapter V, G/M is homeomorphic with $(G/H)/(M/H)$. This completes the proof.

In general, the canonical mapping $\varphi:G \to G/H$ is not closed (§8). For example, $\varphi:R \to R/Z$ (R is the group of the real numbers and Z the group of all integers) is not closed, since the closed set $n + \dfrac{1}{2^n}$ ($n \geq 1$) is mapped onto a nonclosed set $\dfrac{1}{2^n}$ ($n \geq 1$). However, the following is true:

Proposition 21. *Let H be a compact invariant subgroup of a topological group G. Then the canonical mapping* $\varphi:G \to G/H$ *is closed.*

PROOF. Let C be a closed subset of G. Let $\dot{x} \in (G/H) \sim \varphi(C)$, where $\dot{x} = xH$ and $x \notin CH$. Since C is closed and H is compact, by Corollary 3, §20, CH is closed. Therefore, there exists an open set U such that $x \in U \subset G \sim CH$. Since φ is open, $\varphi(U)$ is an open neighborhood of $\dot{x}$ and $\varphi(U) \subset G/H \sim \varphi(C)$. Since $\dot{x}$ is arbitrary, this proves that $G/H \sim \varphi(C)$ is open and hence that $\varphi(C)$ is closed.

Corollary 4. *Let H be an invariant subgroup and M any closed subgroup of a topological group G such that* $H \subset M$. *Then* $\varphi(M)$ *is a closed subgroup of* G/H.

PROOF. Since φ is a homomorphism, $\varphi(M)$ is clearly a subgroup of G/H. Furthermore, by hypothesis, $\varphi^{-1}(\varphi(M)) = MH = M$ which is a closed subgroup of G. Since a subset of a quotient space is closed if, and only if, its inverse image under the canonical mapping is closed (Kelley,[27] p. 94), it follows that $\varphi(M)$ is a closed subgroup of G/H.

25. PRODUCTS AND INVERSE LIMITS OF GROUPS

As in the case of semitopological groups (Chapter II, §14, Proposition 2), the direct product of topological groups is also a topological group. More specifically, we have the following:

Theorem 12. *Let A be an index set. For each* $\alpha \in A$, *let* G_α *be a topological group. Then* $G = \prod_{\alpha \in A} G_\alpha$, *endowed with the product topology, is a topological group.*

PROOF. We have to show only that the mapping: $(x, y) \to xy^{-1}$ of $G \times G$ onto G is continuous. Let W be a neighborhood of xy^{-1} in G. Then there exists a finite set $\{\alpha_i\}$ $(1 \le i \le n)$ of indices in A such that $U \subset W$, where $U = \prod_{\alpha \in A} U_\alpha$, $U_\alpha = G_\alpha$ for all $\alpha \in A$, $\alpha \ne \alpha_i$ $(1 \le i \le n)$, and U_α is an open neighborhood of $x_\alpha y_\alpha^{-1}$ for $\alpha = \alpha_i$ $(1 \le i \le n)$. Since $(x_\alpha, y_\alpha) \to x_\alpha y_\alpha^{-1}$ is continuous for each $\alpha \in A$, there exist neighborhoods V_{α_i}, V'_{α_i} of x_{α_i} and y_{α_i}, respectively, for each α_i $(1 \le i \le n)$ such that $V_{\alpha_i} V'^{-1}_{\alpha_i} \subset U_{\alpha_i}$ for each $i = 1, \ldots, n$. Now let $V = \prod_{\alpha \in A} V_\alpha$, where $V_\alpha = G_\alpha$ for all $\alpha \in A$, $\alpha \ne \alpha_i$ and $V_\alpha = V_{\alpha_i}$ for $\alpha = \alpha_i$. Similarly, $V' = \prod_{\alpha \in A} V'_\alpha$. Then V and V' are neighborhoods of x and y, respectively, and

$$VV'^{-1} = \prod_{\alpha \in A} (V_\alpha V'^{-1}_\alpha) \subset \prod_{\alpha \in A} U_\alpha = U \subset W.$$

This proves the proposition.

The projection mapping $p_\alpha : G \to G_\alpha$ is a continuous and open homomorphism of G onto G_α for each α, as was shown in Proposition 3 (Chapter II, §14).

The following results are useful in the sequel:

Theorem 13. *Let* $G = \prod_{\alpha \in A} G_\alpha$ *be the direct product of topological groups, endowed with the product topology. Then:*

(a) *G is compact if, and only if, each G_α is;*

(b) *G is a Hausdorff topological group if, and only if, each G_α is;*

(c) *G is locally compact if all G_α are compact topological groups except for a finite number of $\alpha_i (1 \le i \le n)$, say, and if for these α_i, G_{α_i} is a locally compact topological group.*

(d) *G is metrizable if A is a countable set of indices and each G_α is a metrizable topological group.*

(e) *G is complete if each G_α is complete.*

PROOF. (a) In view of Theorem 12, G is compact by the Tychonoff theorem. Conversely, if G is compact then $p_\alpha(G) = G_\alpha$ is compact, because p_α is continuous for each α.

(b) It is sufficient to show that G is a T_0-space (§21, Theorem 4). Let $x \ne e$, then at least for some α, $x_\alpha \ne e_\alpha$. Since each G_α is Hausdorff, there exists an open neighborhood U_α of e_α such that $x_\alpha \notin U_\alpha$. Then $p_\alpha^{-1}(U_\alpha)$ is an open neighborhood of e and $x \notin p_\alpha^{-1}(U_\alpha)$. The reverse implication is obvious.

(c) By (a), $H = \prod_{\alpha \ne \alpha_i} G_\alpha$ $(1 \le i \le n)$ is a compact (Theorem 2, §9) topological group and therefore locally compact. Hence, in order to establish (c), one has merely to show that the finite product of locally compact spaces is locally compact. For this, let U_{α_i} be a neighborhood of the identity

e_{α_i} in G_{α_i} such that $\bar{U}_{\alpha_i}$ is compact. Then $U = \prod_{\alpha_i} U_{\alpha_i}$ is a neighborhood of

the identity e in G, and $\bar{U} = \prod_{\alpha_i} \bar{U}_{\alpha_i}$ is compact. This establishes (c).

(d) Let $A = N = \{n : n \geq 1\}$. For each $n \geq 1$, let $U_k^{(n)}$ $(k \geq 1)$ be a
countable fundamental system of neighborhoods of e_n in G_n and satisfying
conditions (a)–(c) of Theorem 3, §20.

Then $U_{nk} = \prod U_k^{(n)}$, where $U_k^{(n)} = G_n$ for $k \neq n$, forms a countable
system of neighborhoods of e in G. Clearly, the family $\{U_{nk}\}(n \geq 1, k \geq 1)$
forms a subbasis of the product topology. Since the base of neighborhoods
of e under the product topology is the family of finite intersections of U_{nk}'s, so
it is countable and satisfies conditions (a)–(c) of Theorem 3, §20. Hence
G is metrizable.

(e) Since G is a uniform space, the result follows from a general theorem
concerning uniform spaces.

Definition 5. (a) Let $(\Gamma, \leq)$ be a directed set (§6, Chapter I). For
each $\alpha \in \Gamma$, let G_α be a topological group and for each pair α, $\beta \in \Gamma$, $\alpha \leq \beta$,
let $f_{\alpha\beta}$ be a continuous homomorphism of G_β into G_α such that $f_{\alpha\alpha} : G_\alpha \to G_\alpha$ is
the identity homomorphism and for $\alpha \leq \beta \leq \gamma$ it follows that $f_{\alpha\gamma} = f_{\alpha\beta} \circ f_{\beta\gamma}$,
i.e., for $x_\gamma \in G_\gamma$, $f_{\alpha\gamma}(x_\gamma) = f_{\alpha\beta}(f_{\beta\gamma}(x_\gamma))$. Then $(G_\alpha, \Gamma, f_{\alpha\beta})$ is called an *inverse
system of topological groups*.

(b) Let $(G_\alpha, \Gamma, f_{\alpha\beta})$ be an inverse system of topological groups. Let
$G = \prod_{\alpha \in \Gamma} G_\alpha$ be the direct product of topological groups. The subset G_∞ of
G consisting of those functions $x = (x_\alpha)$ in G for which $\alpha \leq \beta$ implies
$f_{\alpha\beta}(x_\beta) = x_\alpha$, is called the *inverse limit* or *projective limit* of the inverse system.

Proposition 22. *Let G_∞ be the inverse limit of an inverse system
$(G_\alpha, \Gamma, f_{\alpha\beta})$ of Hausdorff topological groups. Then G_∞ is a closed subgroup of
the product $G = \prod_{\alpha \in \Gamma} G_\alpha$.*

PROOF. Let $x, y \in G_\infty$, $xy^{-1} = (x_\alpha y_\alpha^{-1})$. For $\alpha \leq \beta$, $f_{\alpha\beta} : G_\beta \to G_\alpha$ implies
$f_{\alpha\beta}(x_\beta y_\beta^{-1}) = f_{\alpha\beta}(x_\beta)f_{\alpha\beta}(y_\beta^{-1})$, because $f_{\alpha\beta}$ is a homomorphism. Hence, for
$\alpha \leq \beta$, $x_\alpha y_\alpha^{-1} = f_{\alpha\beta}(x_\beta)f_{\alpha\beta}(y_\beta^{-1}) = f_{\alpha\beta}(x_\beta y_\beta^{-1})$ implies $xy^{-1} \in G_\infty$. This shows
that G_∞ is a subgroup of G. To show that G_∞ is a closed subgroup, let
$x = (x_\alpha) \in G$, $x \notin G_\infty$. Then for some α, $\beta \in \Gamma$, $\alpha \leq \beta$, $x_\alpha \neq f_{\alpha\beta}(x_\beta)$. Since
G_α is Hausdorff, there exist disjoint open sets U_α and V_α such that $x_\alpha \in U_\alpha$,
$f_{\beta\alpha}(x_\beta) \in V_\alpha$. Let $U_\beta = f_{\alpha\beta}^{-1}(V_\alpha)$ which is open in G_β because $f_{\alpha\beta}$ is continuous.
Now let $U = \prod_{\gamma \in \Gamma} U_\gamma$, where $U_\gamma = U_\alpha$ or U_β according as $\gamma = \alpha$ or β, and
$U_\gamma = G_\gamma$ for all $\gamma \neq \alpha$, β. Then clearly U is an open neighborhood of x and
U does not meet G_∞. This completes the proof.

Corollary 5. *Let G_∞ be the inverse limit of an inverse system $(G_\alpha, \Gamma, f_{\alpha\beta})$
of Hausdorff compact topological groups. Then G_∞ is a compact Hausdorff
topological subgroup.*

PROOF. By Proposition 22, G_∞ is a closed Hausdorff topological subgroup of the product $G = \prod_{\alpha \in \Gamma} G_\alpha$. Since G is compact (Theorem 13 (a)), so is G_∞.

Exercises

1. Show that topological spaces described in Examples (§19) are topological groups.

2. Show that T^n (§19, Example 6) is a compact topological group.

3. Let $\{U\}$ be the system of neighborhoods of the identity e in a topological group G. Show that for any subset A in G,

$$\bar{A} = \bigcap UAU.$$

4. For a subgroup H of a topological group G to be closed, it is necessary and sufficient that there exists an open subset U of G such that $U \cap H \neq \varnothing$ and $U \cap H = U \cap \bar{H}$.

5. For a subgroup H of a topological group G to be discrete, it is necessary and sufficient that H has an isolated point. (A point of H is said to be *isolated* if it is both open and closed in H.)

6. Let G be a topological group and H a subgroup of G. If H and G/H both are compact, so is G.

7. (a) If G is a connected (or locally connected) topological group and H a subgroup of G, then G/H is also connected (or locally connected).

(b) Let G be a topological group and H a subgroup of G. If H and G/H are connected, so is G.

(c) Let G be a topological group. Show that $\overline{\{e\}}$ is an invariant closed subgroup of G and, hence, $G/\overline{\{e\}}$ is a Hausdorff topological group.

8. Let $G = \prod_{\alpha \in A} G_\alpha$ be the direct product of connected topological groups $G_\alpha (\alpha \in A)$, where G is endowed with the product topology. Show that G is connected.

9. Let $G = \prod_{\alpha \in A} G_\alpha$, where G_α is a topological group for each α. Let G_b be endowed with the *box topology* defined as follows: The collection of sets $\prod_{\alpha \in A} U_\alpha$ where, for each $\alpha \in A$, U_α is an open set of G_α, forms the subbase of the box topology. Show that G_b is a topological group and the topology b is finer than the product topology.

10. Let G be any Hausdorff topological abelian group. Show that there exists a complete Hausdorff topological group $\hat{G}$ such that G is homeomorphically dense in $\hat{G}$.

11. Let R be the additive group of real numbers. Let G denote the set of all affine transformations of R into R, i.e., $f \in G$ if $f(x) = ax + b$, where $a \neq 0$. Show that G is a group under composition. Define $U_{\varepsilon, \eta}(e) = \{f : f(x) = ax + b, |a - 1| < \varepsilon, |b| < \eta; \varepsilon, \eta > 0\}$ as the system of neighborhoods of the identity transformation $e \in G$, when ε, η run over the set of

positive real numbers. Then G is a topological group. Show that for this group the left and right uniformities are distinct.

12. Two topological groups G and H are said to be *locally isomorphic* if there exist neighborhoods U and V of the identities of G and H respectively and a homeomorphism f of U onto V such that:

(i) for each pair x, $y \in U$ such that $xy \in U$, $f(xy) = f(x)f(y)$;

(ii) if g is the inverse of f, then for each pair, x', $y' \in V$ such that $x'y' \in V$, $g(x'y') = g(x')g(y')$.

(a) If G is a topological group and H a discrete invariant subgroup of G, then G and G/H are locally isomorphic. In particular, R and $T = R/Z$ (Examples 1 and 4, §19) are locally isomorphic. This shows that sufficiently small neighborhoods of identities in R and T can be identified.

(b) Show that for any two connected locally isomorphic topological groups G and H, there exists a group K such that G and H are isomorphic with K/N and K/M, respectively, for some discrete invariant subgroups N and M of K (cf., Pontrjagin[37]).

(c) Let m and n be two positive integers such that $m < n$. Let the Euclidean space R^m be embedded into R^n. Let Z^m denote the subgroup of R^m with integral coordinates. Let the quotient group R^n/Z^m be denoted by T_m^n.

Show that all topological groups T_m^n $(1 \leq m \leq n)$ are locally isomorphic with each other, but not isomorphic.

13. Show that each proper subgroup of the additive group of real numbers is closed if, and only if, it is discrete.

14. Let f be a continuous homomorphism of any subgroup F of the additive group R of real numbers into a locally compact topological group G. Then either f is a homeomorphism onto $f(F)$, or $\overline{f(F)}$ is a compact abelian subgroup of G. If f is not a homeomorphism, then for each neighborhood U of e in G, there exists $\alpha > 0$ in R such that $f\{[x, x + \alpha) \cap F\} \cap U \neq \varnothing$, for every $x \in F$, where $[x, x + \alpha)$ is a left-closed and right-open interval.

IV

Locally Compact Groups

In this chapter, we study the general properties of locally compact groups. We show that a locally compact Hausdorff topological group is normal. (Not all topological groups are normal; e.g., §21, Chapter III.) As particular cases of locally compact groups the classical groups, viz., full linear groups, groups of unitary and orthogonal matrices, etc., are discussed. Furthermore, we indicate that these classical groups are Lie groups.

26. GENERAL RESULTS ON LOCALLY COMPACT GROUPS

As we saw in Chapter III (Theorem 5, §21), every Hausdorff topological group is completely regular. Also there exists a topological group which is not normal (Example, §21). Indeed, every Hausdorff compact topological space (in particular, topological group) is normal. We show that the same is true for locally compact Hausdorff topological groups.

Proposition 1. *Every topological group is locally compact if, and only if, there exists a compact neighborhood of the identity.*

PROOF. Let G be a locally compact topological group. Then there exists a neighborhood U of e such that $\bar{U}$ is compact. Conversely, if there exists a compact neighborhood U of e, then there exists a neighborhood V of e such that $V^2 \subset U$. But $\bar{V} \subset V^2 \subset U$. Hence $\bar{V}$, being a closed subset of a compact set $\bar{U}$, is compact and is also a neighborhood of e. Now for each $x \in G$, xV is a neighborhood of x and $\overline{xV} = x\bar{V}$ is compact because the translations are homeomorphisms.

Theorem 1. *Every locally compact Hausdorff topological group is normal.*

PROOF. In a locally compact Hausdorff topological group G let U be a symmetric neighborhood of e and let $\bar{U}$ be compact. Consider $H = \bigcup_{n \geq 1} U^n$. Then H is an open and closed subgroup of G (Proposition 10, §23, Chapter III). Hence, by Theorem 7, §23, Chapter III, H is a Hausdorff locally compact subgroup of G. Furthermore, by Proposition 3, §20, Chapter III, $\bar{U}^{n-1} \subset U^{n-1}U = U^n$ for $n \geq 1$. Hence, $H = \bigcup_{n \geq 1} U^n = \bigcup_{n \geq 1} \bar{U}^n$. But $\bar{U}^n$ is compact because $\bar{U}$ is compact, whence H is the union of an increasing sequence of compact sets. Therefore, H is normal by Proposition 5, §7, Chapter I. Now consider the family of pairwise disjoint cosets $\{aH\}$ in G. Since the right and left translations in G are homeomorphisms, each aH, being homeomorphic with H, is normal. Since $G = \bigcup aH$, G is normal by Proposition 4, §5, Chapter I.

Corollary 1. *Let G be a locally compact Hausdorff topological group, C a closed (in particular a compact) subset and U an open set containing C. Then there exists a continuous real valued function f such that $f(C) = 1$ and $f(G \sim U) = 0$.*

PROOF. Since G is normal by the above theorem, and since C and $G \sim U$ are closed disjoint subsets of G, the result follows by Urysohn's lemma.

Theorem 2. *Every locally compact Hausdorff topological group G satisfying the second axiom of countability is metrizable (Definition 2, §21, Chapter III).*

PROOF. By Theorem 16 (Kelley,[27] p. 125), G is homeomorphic with a subspace of the Hilbert cube H^ω. Since each point in H^ω has a countable fundamental system of neighborhoods, so does e in G. Since G is Hausdorff, this proves that G is metrizable.

Theorem 3. *Every locally compact Hausdorff topological group is complete in its usual uniform structure.*

PROOF. Let G be the locally compact Hausdorff topological group with its right uniform structure. Let $\mathscr{K}$ be a Cauchy filter and U a compact neighborhood of the identity $e \in G$. There exists a set A in $\mathscr{K}$ such that for all x, $y \in A$, $yx^{-1} \in U$. That means $A \subset Ux$. Since U is compact and the translations continuous, Ux is compact. Clearly, $\mathscr{K} \cap Ux$ is a filter on Ux and it has a limit point x_0 in Ux. Therefore, x_0 is a limit point of $\mathscr{K}$. Since $\mathscr{K}$ is a Cauchy filter and G a Hausdorff space, $\mathscr{K}$ converges to x_0.

Corollary 2. *Every Hausdorff locally compact topological group G satisfying the second axiom of countability is metrizable and complete.*

PROOF. By Theorems 2 and 3, *G* is metrizable and complete.

Corollary 3. *Every locally compact subgroup of a topological group is a closed subgroup.*

PROOF. In view of Theorem 3, this follows from the fact that every complete subset of a uniform space is closed (Proposition 7, §10, Chapter I).

Theorem 4. *Let E be a locally compact topological group and F any Hausdorff topological group. Let f be a continuous and almost open homomorphism of E into F. Then F is also locally compact.*

PROOF. In view of Proposition 1, it is sufficient to show that the identity of *F* has a compact neighborhood. Let *U* be a compact neighborhood of the identity in *E*. Then $f(U)$ is a compact closed subset of *F*, since *f* is continuous, *U* is compact and *F* is Hausdorff. But, *f* being an almost open homomorphism, $\overline{f(U)} = f(U)$ is a neighborhood of the identity in *F*. This proves the theorem.

Corollary 4. *Let E be a locally compact topological group and F any topological group. Let f be a continuous and open homomorphism of E into F. Then F is also a locally compact topological group.*

PROOF. For each compact open neighborhood *U* of *e* in *E*, it follows, as in the theorem above, that $f(U)$ is a compact open neighborhood of the identity in *F*. Hence, by Proposition 1, *F* is locally compact.

If, in addition, *F* is a Hausdorff space, Corollary 4 is a particular case of Theorem 4, since each open mapping is almost open.

For any invariant subgroup *H* of a locally compact topological group *G*, the canonical mapping $\varphi: G \to G/H$ is a continuous and open homomorphism. Thus, one derives from Corollary 4 that *G/H* is locally compact, as was shown directly in Theorem 11, §24, Chapter III for Hausdorff groups *G* if *H* is a closed invariant subgroup of *G*. A sort of converse is also true. For a more precise statement, see Exercise 2.

Proposition 2. *Let G be a Hausdorff topological group, M a locally compact subgroup of G, and H a closed invariant subgroup of G. Define the mapping: $f(\dot{m}) = m(M \cap H)$, $(m \in M)$, of MH/H onto M/(M ∩ H) as in Proposition 19, §24, Chapter III. If f is almost continuous, then MH is closed in G.*

PROOF. Since *M* is locally compact and $M \cap H$ is a closed invariant subgroup of *M*, $M/(M \cap H)$ is also locally compact by the remark after

Corollary 4. But then f, being an algebraic isomorphism of MH/H, onto $M/(M \cap H)$ has its inverse $f^{-1}: M/(M \cap H) \to MH/H$, which is almost open because f is almost continuous by hypothesis. Moreover, f^{-1} is continuous because f is open by Proposition 19, §24, Chapter III. Since H is a closed invariant subgroup, MH/H is a Hausdorff topological group. Hence, by Theorem 4, MH/H is locally compact. But then by Corollary 3, MH/H is a closed subgroup of G/H. Hence, its inverse image MH in G under the continuous canonical mapping $\varphi: G \to G/H$, is also closed.

Proposition 3. *Let G be a locally compact topological group. Let C be a compact subset and U an open subset of G such that $C \subset U$. Then there exists a neighborhood V of e such that $\overline{(CV \cup VC)}$ is compact and $\overline{CV \cup VC} \subset U$.*

PROOF. For each $x \in C$, there exists an open neighborhood V_x of the identity e such that $xV_x \subset U$. Also for each V_x there exists an open neighborhood W_x of e such that $W_x^2 \subset V_x$. Clearly, $\{xW_x\}$ is an open covering of C. Since C is compact, the union of a finite number of sets $\{x_iW_{x_i}\}(1 \leq i \leq n)$ covers C. Let $W_1 = \bigcap_{i=1}^{n} W_{x_i}$. Then W_1 is an open neighborhood of e. Thus,

$$CW_1 \subset \bigcup_{i=1}^{n} x_iW_{x_i}W_1 \subset \bigcup_{i=1}^{n} x_iW_{x_i}^2 \subset \bigcup_{i=1}^{n} x_iV_{x_i} \subset U.$$

Similarly, there exists a neighborhood W_2 of e such that $W_2C \subset U$. Since $W_1 \cap W_2$ is a neighborhood of e and G is locally compact, there exists a neighborhood V of e such that $\bar{V}$ is compact and $\bar{V} \subset W_1 \cap W_2$. Hence, $C\bar{V} \cup \bar{V}C \subset U$. Since $\bar{V}$ is closed and C compact, each of $C\bar{V}$ and $\bar{V}C$ is a closed set by Corollary 3, §20, Chapter III. But $C\bar{V} \cup \bar{V}C$, as the union of two compact sets (since each of $C\bar{V}$ and $\bar{V}C$, as the product of two compact sets, is compact), is compact. Also $\overline{CV} = C\bar{V}$ and $\overline{VC} = \bar{V}C$. Hence $\overline{CV \cup VC} = C\bar{V} \cup \bar{V}C$ is compact. Furthermore, $\overline{CV \cup VC} = C\bar{V} \cup \bar{V}C \subset U$. This proves the proposition.

Theorem 5. *Every compact open neighborhood U of the identity of a topological group G contains a compact, open, and closed subgroup.*

PROOF. By the above proposition, there exists an open neighborhood V of e such that $UV \subset U$. Since $U \cap V$ is a neighborhood of e, there exists a symmetric neighborhood W of e such that $W \subset U \cap V$. Thus, $W^2 \subset UV \subset U$. But then by induction, $W^n = W^{n-1}W \subset UW \subset UV \subset U$ for each $n \geq 1$. Hence, $H = \bigcup_{n \geq 1} W^n \subset U$. By Proposition 10, §23, Chapter III, H is an open and closed subgroup of G. Since H is a closed subset of a compact set U, H is also compact.

Proposition 4. *Let U be any open neighborhood of e in a topological group G, and C any compact subset of G. Then there exists an open neighborhood V of e such that $CVC^{-1} \subset U$.*

PROOF. By Proposition 2, §20, Chapter III, there exists a symmetric open neighborhood W_1 of e such that $W_1^3 \subset U$, and a symmetric open neighborhood W_2 of e such that $aW_2a^{-1} \subset W$ for a fixed $a \in G$ (Theorem 2, §19, Chapter III). Let $W = W_1 \cap W_2$. Then for each $x \in Wa$, we have $xa^{-1} \in W \subset W_1$ and so $(xa^{-1})^{-1} = ax^{-1} \in W_1^{-1} = W_1$. Hence, $xWx^{-1} \subset xW_2x^{-1} = (xa^{-1})aW_2a^{-1}(ax^{-1}) \subset W_1^3 \subset U$.

Thus, it follows that for each $a \in C$, there exists an open symmetric neighborhood W of e such that $xWx^{-1} \subset U$ whenever $x \in Wa$. Since W depends upon a, we shall designate W by W_a. As a runs over C, $\{W_a a\}$ is an open covering of C. Since C is compact, the union of a finite number of sets $\{W_{a_i} a_i\}(1 \le i \le n)$ covers C. Let $V = \bigcap_{i=1}^{n} W_{a_i}$. Then V is an open symmetric neighborhood of e. Now for each $x \in C$, there exists k $(1 \le k \le n)$ such that $x \in W_{a_k} a_k$, which implies $xW_{a_k}x^{-1} \subset U$. Hence, $xVx^{-1} \subset xW_{a_k}x^{-1} \subset U$. This completes the proof.

Theorem 6. *Each closed and open neighborhood U of e in a compact topological group G contains an open and closed invariant subgroup M of G. Also G/M is finite.*

PROOF. Since G is compact and U is closed, U is also compact. Therefore, by Theorem 5, there exists a compact open and closed subgroup H of G such that $H \subset U$. Let $M = \bigcap_{x \in G} xHx^{-1}$. Then M is a closed invariant subgroup of G and $M \subset U$. By Proposition 4, there exists an open neighborhood V of e such that $x^{-1}Vx \subset H$ for all $x \in G$, because G is compact and H, being an open subgroup, is a neighborhood of e. Therefore, $V \subset xHx^{-1}$ for all $x \in G$. Hence, $V \subset M$ and so M is also open. Therefore, G/M is a discrete compact group by Proposition 16 and Theorem 11 of §24, Chapter III. But a compact discrete space is finite. This proves the theorem.

Theorem 7. *Every neighborhood of e in a totally disconnected locally compact (in particular, compact) topological group contains a compact open (compact, open and invariant) subgroup.*

PROOF. Since every neighborhood U of e in G contains a compact or compact closed neighborhood of e, Theorem 7 follows from Theorems 5 and 6.

Proposition 5. *The component C of the identity e of a locally compact topological group G is the intersection of all open subgroups of G.*

PROOF.   Let H be any open subgroup of G. Then H is closed (Proposition 7, §23, Chapter III).   If $C \not\subseteq H$, $C \cap H$ would be a closed and open proper subset of a connected set C, which is not possible.   Hence, $C \subseteq H$ because $e \in H$.   Therefore, $C \subseteq \cap H$, where intersection is taken over all open subgroups of G.   On the other hand, let $x \notin C$.   Since G/C is totally disconnected (Proposition 17, §24, Chapter III) and a locally compact Hausdorff topological group (Theorem 11, §24, Chapter III), by Theorem 7 there exists a compact open subgroup $\dot{H} = HC/C$ of G/C such that $\dot{x} \notin \dot{H}$. Since $\dot{H}$ is open in G/C, HC is an open subgroup of G and $x \notin HC$.   This shows that $C = \cap H$.

For locally compact groups, Theorem 6 of Chapter III, §23, can be improved as follows:

Theorem 8.   *Let G be a locally compact topological group.   The following statements are equivalent:*
  (a) *G is connected;*
  (b) *G has no proper open subgroups;*
  (c) *For each neighborhood U of e, $\bigcup\limits_{n \geq 1} U^n = G$.*

PROOF.   Since each open subgroup is closed, (a) $\Rightarrow$ (b).   Clearly, (b) $\Rightarrow$ (c).   Since the component C of the identity is contained in each open subgroup of G (Proposition 5), $C = G$.   This establishes the implication (c) $\Rightarrow$ (a).

27. CLASSICAL LINEAR GROUPS

In this section, we study a subclass of locally compact groups, viz., groups of matrices.   Since the study of general topological groups is very recent in comparison with that of matrices, the customary word "classical" is used for the groups of matrices.   The word "linear" is taken from linear algebra, in which the study of matrices is the main theme.

Let K denote the field R or C of real or complex numbers.   For a positive integer n, K^n denotes the direct product of K by itself n-times, i.e., $x \in K^n$ if, and only if, $x = (x_1, \ldots, x_n)$, where $x_i \in K$, $1 \leq i \leq n$.   With the usual definition of addition and scalar multiplication, K^n is a vector over the field K.   More specifically, the addition and scalar multiplication are defined as follows: For each pair $x, y \in K^n$,

$$x + y = (x_1 + y_1, \ldots, x_n + y_n), \text{ where } x = (x_1, x_2, \ldots, x_n) \text{ and }$$
$$y = (y_1, y_2, \ldots, y_n); \text{ and for any } \lambda \in K \text{ and } x \in K^n,$$
$$\lambda x = (\lambda x_1, \lambda x_2, \ldots, \lambda x_n).$$

Also, K^n is an algebra if we define $xy = (x_1 y_1, \ldots, x_n y_n)$.

Let $e_i = (0, \ldots, 1, \ldots, 0)$, where 1 occurs at ith place and 0 elsewhere. Then it is known that the set $\{e_i\}$ $(1 \leq i \leq n)$ forms a basis of K^n.

A mapping f of K^n into itself is said to be an *endomorphism* of K^n if:

(i) $f(x + y) = f(x) + f(y)$, $x, y \in K^n$;

(ii) $f(\lambda x) = \lambda f(x)$, $\lambda \in K$.

It is convenient to replace the endomorphisms of K^n by certain objects called matrices. For this we have the following:

Let a_{ij}, $1 \le i \le m$, $1 \le j \le n$, be mn elements of K. The array

$$\begin{bmatrix} a_{11}, & a_{12}, & \ldots, & a_{1n} \\ a_{21}, & a_{22}, & \ldots, & a_{2n} \\ \ldots, & \ldots, & \ldots, & \ldots \\ \ldots, & \ldots, & \ldots, & \ldots \\ a_{m1}, & a_{m2}, & \ldots, & a_{mn} \end{bmatrix}$$

is called a *matrix* of m *rows* and n *columns*, and is denoted briefly by (a_{ij}), $1 \le i \le m$, $1 \le j \le n$. A matrix of m rows and n columns is usually called a matrix of *order* $m \times n$. If $m = n$ (i.e., the number of rows equals the number of columns) the matrix (a_{ij}), $1 \le i \le m$, $1 \le j \le n$, is called a *square* matrix of order $m \times m$ or $n \times n$. The elements a_{ij} are called *entries* or, *coefficients* of the matrix. $[a_{i1}, a_{i2}, \ldots, a_{in}]$ is called the ith row and $\{a_{1j}, a_{2j}, \ldots, a_{nj}\}$ is called the jth column of the matrix.

Let $A = (a_{ij})$ and $B = (b_{ij})$, $1 \le i \le m$, $1 \le j \le n$ be two matrices of order $m \times n$. The *sum* $A + B$ of A and B is defined to be a matrix of order $m \times n$ with entries c_{ij}, $1 \le i \le m$, $1 \le j \le n$, where $c_{ij} = a_{ij} + b_{ij}$. The product of matrices A and B of order $m \times n$ is not always defined. If the number of columns of A equals the number of rows of B, then the product AB of A and B is defined as follows: Let $A = (a_{ij})$, $1 \le i \le m$, $1 \le j \le n$ and $B = (b_{ij})$, $1 \le i \le n$, $1 \le j \le l$. Then $AB = (d_{ij})$, where $d_{ij} = \sum_{k=1}^{n} a_{ik} b_{kj}$; $1 \le i \le m$, $1 \le i \le l$ is the product matrix of order $m \times l$. Although $A + B = B + A$, BA and BA need not coincide (even when both are defined!). In other words, addition is commutative but multiplication is not. Both addition and multiplication are associative.

In particular, if $x = [x_1, x_2, \ldots, x_n]$ is a matrix of order $1 \times n$ and $A = (a_{ij})$, $1 \le i \le n$, $1 \le j \le n$, then Ax is not defined. However, xA is a matrix of order $1 \times n$. Let $xA = [y_1, y_2, \ldots, y_n]$. Then $y_i = \sum_{k=1}^{n} a_{ki} x_k$. If

we denote x by $\{x_1, x_2, \ldots, x_n\} = \begin{bmatrix} x_1 \\ x_2 \\ \cdot \\ \cdot \\ \cdot \\ x_n \end{bmatrix}$ a matrix of order $n \times 1$, then Ax

is defined and is a matrix of order $n \times 1$.

If a matrix of order $n \times 1$ as well as a matrix of order $1 \times n$ is identified with a vector or an element in K^n, then xA and Ax are elements of K^n. In other words, the action of A on $x \in K^n$ can be regarded as a mapping of K^n into itself, or as an endomorphism of K^n. Clearly, $(x + y)A = xA + yA$, for $x, y \in K^n$.

As in the case of scalar multiplication in K^n, scalar multiplication for matrices is performed entrywise, viz., if $A = (a_{ij})$, $1 \leq i, j \leq n$ and $\lambda \in K$, then $\lambda A = (\lambda a_{ij})$, $1 \leq i, j \leq n$. Thus, it is clear that the action of a square matrix of order $n \times n$ on the elements of K^n is linear. Proposition 7 below establishes a relation between the set of square matrices of order $n \times n$ and the set of endomorphisms of K^n.

With square matrices one can associate the concept of *determinant*. We shall not enter into an abstract definition of a determinant. We appeal to the principle of induction on the order of matrices. Let $A = (a_{ij})$ $1 \leq i$, $j \leq 2$ be a square matrix of order 2×2. Then the *determinant* $|A|$ of A is equal to $a_{11}a_{22} - a_{21}a_{12}$. The determinant of a square matrix of order 3×3 can be expressed in terms of determinants of matrices of order 2×2, as follows: Let $A = (a_{ij})$, $1 \leq i, j \leq 3$. Then

$$|A| = a_{11} \begin{vmatrix} a_{22}, & a_{23} \\ a_{32}, & a_{33} \end{vmatrix} - a_{21} \begin{vmatrix} a_{12} & a_{13} \\ a_{32} & a_{33} \end{vmatrix} + a_{31} \begin{vmatrix} a_{12} & a_{13} \\ a_{22} & a_{23} \end{vmatrix}.$$

The determinants of order 2 involved in expressing $|A|$ are called the *minors* of $|A|$. By a familiar induction we may define the concept of a determinant for a square matrix of arbitrary order $n \times n$.

If A is a square matrix of order $n \times n$ with entries in K, clearly $|A| \in K$. If $A = (a_{ij})$, $1 \leq i, j \leq n$ is a square matrix such that $a_{ij} = 0$ for $i \neq j$ and $a_{ij} = 1$ for $i = j$, then A is called the *identity* matrix of order $n \times n$, and is denoted by I. Clearly $|I| = 1$. If A and B are square matrices of order $n \times n$, it can be shown that $|AB| = |A| \, |B|$. If A is a square matrix of order $n \times n$, a solution of the polynomial $|A - \lambda I| = 0$ is called a *characteristic root* or *eigen value* of A.

Proposition 6. *The set E of all endomorphisms of K^n forms an algebra.*

PROOF. Let f, g be two endomorphisms of K^n. Define $f + g = h$ so that $h(x) = f(x) + g(x)$ for each $x \in K^n$. Then clearly h is an endomorphism of K^n. If 0 denotes the zero-endomorphism, i.e., $0(x) = 0 \in K^n$ for each $x \in K^n$ and $(-f)(x) = -f(x)$, then it is easily seen that E is an abelian additive group. Furthermore, if, for $\lambda \in K$, $\lambda f(x) = \lambda(f(x))$ for each $x \in K^n$, then it follows that E is even a vector space. For each pair $f, g \in E$, define $fg(x) = f(g(x))$. Then with the composition as the product of elements in E, one verifies that E is a commutative algebra.

Proposition 7. *To each endomorphism of K^n there corresponds a matrix of order $n \times n$. Conversely, a matrix A of order $n \times n$ induces an endomorphism of K^n.*

PROOF. Let $e_i = (0, \ldots, 1, \ldots, 0)$, where 1 occurs at the ith coordinate and 0 elsewhere. Then e_i ($1 \leq i \leq n$) is a basis of K^n. Let f be an endomorphism of K^n. Let $f(e_i) = \sum_{j=1}^{n} a_{ij}e_j$, $1 \leq i \leq n$. Then $A = (a_{ij})$ is a matrix of order $n \times n$.

Further, if $x \in K^n$, then $x = \sum_{i=1}^{n} \lambda_i e_i = (\lambda_1, \ldots, \lambda_n)$, $\lambda_i \in K$. Hence,

$$f(x) = \sum_{i=1}^{n} \lambda_i f(e_i) = \sum_{i=1}^{n} \lambda_i \sum_{j=1}^{n} a_{ij}e_j = \sum_{j=1}^{n} \left(\sum_{i=1}^{n} \lambda_i a_{ij} \right) e_j = \left(\sum_{i=1}^{n} \lambda_i a_{i1}, \ldots, \sum_{i=1}^{n} \lambda_i a_{in} \right) =$$

$(\lambda_1, \ldots, \lambda_n)A = xA$. Since the action of A on the elements of K^n is linear, we see that the endomorphism f can be represented by the action of A. Conversely, if $A = (a_{ij})$ is a square matrix of order n, then for each $x = (x_1, \ldots, x_n) \in K^n$, xA is a vector with n coordinates each in K. Therefore, $f(x) = xA \in K^n$ is an endomorphism of K^n, as is easy to check.

Notation 1. $\mathbf{M}_n(K)$ denotes the set of all $n \times n$ matrices with coefficients in K.

As follows from Propositions 6 and 7, $\mathbf{M}_n(K)$ is an additive abelian group.

Proposition 8. *$\mathbf{M}_n(K)$ is a Hausdorff topological group which is locally compact, noncompact, and satisfies the second axiom of countability.*

PROOF. First of all, we shall assign a topology on $\mathbf{M}_n(K)$ by identifying this with K^m for some m. Let the coefficients of each matrix $A = (a_{ij})$ in $\mathbf{M}_n(K)$ be arranged in a definite order. In that case, A can be regarded as a point in K^{n^2}. Let f be the mapping that, to each A in $\mathbf{M}_n(K)$, assigns the point of K^{n^2} obtained by arranging the coefficients of A in the fixed order. Then it is easy to see that f is $1:1$ and onto. Let $\mathbf{M}_n(K)$ be endowed with the same topology as that of K^{n^2}. In other words, a set T in $\mathbf{M}_n(K)$ is open if, and only if, $f(T)$ is an open subset of K^{n^2} under the usual Euclidean topology. Since K^{n^2} is an additive abelian topological group (Example 5, §19, Chapter III) satisfying all the properties mentioned in the Proposition, so is $\mathbf{M}_n(K)$.

The identification established in Proposition 8, between $\mathbf{M}_n(K)$ and K^{n^2}, helps in the study of groups of matrices since the Euclidean spaces K^{n^2} are somewhat understood.

Definition 1. Let $A = (a_{ij})$ be a matrix of order $n \times n$.

(i) $A' = (a'_{ij}) = (a_{ji})$ is called the *transpose* matrix of A.

(ii) $\bar{A} = \overline{(a_{ij})}$ is called the *complex conjugate*, where $\bar{a}_{ij}$ is the conjugate complex number of a_{ij}.

(iii) A is called *regular* or *nonsingular* if there exists a matrix A^{-1} of order $n \times n$ such that $AA^{-1} = A^{-1}A = I$, the identity matrix. The matrix A^{-1} is called the *inverse* matrix of A.

(iv) For each square matrix A, $|A|$ denotes the *determinant* of A.

The following results are either trivial or well-known from linear algebra:

(a) $A = (a_{ij})$ is real if and only if $A = \bar{A}$.

(b) $(AB)' = B'A'$.

(c) $\overline{AB} = \bar{A}\bar{B}$.

(d) For regular matrices A and B, AB and BA are regular; and $(AB)^{-1} = B^{-1}A^{-1}$.

(e) A is regular if, and only if, $|A| \neq 0$.

Notation 2.   The set of all regular $n \times n$ matrices with coefficients in K is denoted by $\mathbf{G}_n(K)$, *and is called the* general *or* full linear group.

Proposition 9. $\mathbf{G}_n(K)$ *is an open subset of* $\mathbf{M}_n(K)$.

PROOF.   Let $f(A) = |A|$ be the mapping of $\mathbf{M}_n(K)$ into K. Then $\mathbf{G}_n(K) = \mathbf{M}_n(K) \sim f^{-1}(0)$, owing to (e) in the paragraph following the above definition. Since f is a multilinear function, it is continuous. Therefore, $f^{-1}(0)$ is a closed subset of $\mathbf{M}_n(K)$, and hence $\mathbf{G}_n(K)$ is open.

Theorem 9. $\mathbf{G}_n(K)$ *is a Hausdorff multiplicative topological group when it is endowed with the relative topology induced from* $\mathbf{M}_n(K)$.

PROOF.   Indeed, the product of two regular matrices is regular. Moreover, for each $A \in \mathbf{G}_n(K)$, A^{-1} exists and $A^{-1} \in \mathbf{G}_n(K)$. Thus, it is clear that $\mathbf{G}_n(K)$ is a multiplicative group with its identity as I. Since $\mathbf{M}_n(K)$ is a Hausdorff space, so is $\mathbf{G}_n(K)$ under the topology induced from $\mathbf{M}_n(K)$. Now to show that the mapping:

$$(A, B) \to AB$$

is continuous in A and B together, we observe that $AB = (c_{ij})$, where $c_{ij} = \sum_{k=1}^{n} a_{ik}b_{kj}$, $A = (a_{ij})$ and $B = (b_{ij})$. Since the mappings: $A \to a_{ij}$ and $B \to b_{ij}$ are continuous because they are projections of K^{n^2}, so the mapping: $AB \to c_{ij}$ is also continuous. Now the fact that a mapping of a topological space into a space of matrices or, in other words, into a product space K^{n^2} is continuous if, and only if, each coefficient is continuous (§9, Chapter I), establishes the continuity of $(A, B) \to AB$.

Furthermore, if $A \in \mathbf{G}_n(K)$, $A^{-1} = |A|^{-1}(d_{ij})$, where d_{ij}'s are minors of $|A|$ or polynomials in the coefficients of A. Since $|A| \neq 0$, $A \to |A|^{-1}(d_{ij})$ is continuous and so is $A \to A^{-1}$. This proves that $\mathbf{G}_n(K)$ is a Hausdorff topological multiplicative group.

Theorem 10. *The mappings:* $A \to A^{-1}$, $A \to \bar{A}$, $A \to A'$ *and* $A \to A'^{-1} = A^*$ *are homeomorphisms of* $\mathbf{G}_n(K)$ *onto itself. The second and fourth mappings are also automorphisms.*

PROOF. By Theorem 2, §19, Chapter III, $A \to A^{-1}$ is a homeomorphism. For the second mapping, let $f(A) = \bar{A}$. Since for any complex number α, $\bar{\bar{\alpha}} = \alpha$, it is easy to verify that f is $1:1$ and onto. Since $A \to \overline{a_{ij}}$ is continuous where $A = (a_{ij})$, so is f. Further, $f(\bar{A}) = \bar{\bar{A}} = A$, or $f^{-1}(A) = \bar{A}$ shows that f^{-1} is continuous. Hence, f is a homeomorphism. Using similar arguments one shows that the third mapping is also a homeomorphism. The fourth mapping, being the composition of the first and third homeomorphism, is also a homeomorphism. Furthermore, since for any A, $B \in \mathbf{G}_n(K)$, $\overline{AB} = \bar{A}\bar{B}$ and $(AB)^* = A^*B^*$, the second and fourth mappings are automorphisms.

Definition 2. Let A be a matrix of order $n \times n$.
(a) A is *orthogonal* if, and only if, $A = \bar{A} = A^*$.
(b) A is *complex orthogonal* if, and only if, $A = A^*$.
(c) A is *unitary* if, and only if, $\bar{A} = A^* = A'^{-1}$.

Notation 3. The sets of matrices in $\mathbf{G}_n(K)$ satisfying (a), (b) and (c), above will be denoted by $\mathbf{O}_n(R)$, $\mathbf{O}_n(C)$ and $\mathbf{U}_n$, respectively, *and are called the* orthogonal, complex orthogonal, *and* unitary groups of matrices, *respectively.*

As is well-known from linear algebra, the following statements are equivalent:
    (i) A is unitary.
    (ii) $A'\bar{A} = \bar{A}A' = I$.
    (iii) $\bar{A}' = A^{-1}$.
    (iv) $\bar{A}^{-1} = A'$.
    (v) $F(xA) = F(x)$, where $F(x) = \sum\limits_{i=1}^{n} x_i \bar{x}_i$, $x = (x_1, \ldots, x_n) \in C^n$.

Proposition 10. $\mathbf{O}_n(R)$, $\mathbf{O}_n(C)$ *and* $\mathbf{U}_n$ *are topological subgroups of* $\mathbf{G}_n(C)$. *Also* $\mathbf{O}_n(R) = \mathbf{O}_n(C) \cap \mathbf{U}_n$ *and* $\mathbf{O}_n(R) = \mathbf{G}_n(R) \cap \mathbf{O}_n(C)$.

PROOF. By the properties of matrices, given in the paragraph following Definition 1, it is easy to check that $\mathbf{O}_n(R)$, $\mathbf{O}_n(C)$, and $\mathbf{U}_n$ are multiplicative groups. Since $\mathbf{O}_n(R) \subset \mathbf{O}_n(C) \subset \mathbf{G}_n(C)$ and $\mathbf{U}_n \subset \mathbf{G}_n(C)$, the first part of the proposition is immediate.

Since $\mathbf{O}_n(R) \subset \mathbf{O}_n(C)$ and $\mathbf{O}_n(R) \subset \mathbf{U}_n$, $\mathbf{O}_n(R) \subset \mathbf{O}_n(C) \cap \mathbf{U}_n$. Also by Definition 2(a) and (c), the reverse inclusion follows. Hence, the first equality; the second follows similarly.

Proposition 11. $\mathbf{O}_n(R)$, $\mathbf{O}_n(C)$ *and* $\mathbf{U}_n$ *are closed subsets of* $\mathbf{M}_n(C)$. *Hence, they are closed subgroups of* $\mathbf{G}_n(C)$.

PROOF. Since the mappings: $A \to \bar{A}$ and $A \to A^*$ of $\mathbf{M}_n(C)$ onto $\mathbf{M}_n(C)$ are continuous (Theorem 10), the sets $\{A \in \mathbf{M}_n(C): A = \bar{A} = A^*\} = \mathbf{O}_n(R)$, $\{A \in \mathbf{M}_n(C): A = A^*\} = \mathbf{O}_n(C)$, and $\{A \in \mathbf{M}_n(C): \bar{A} = A^*\} = \mathbf{U}_n$ are closed in $\mathbf{M}_n(C)$. Hence, they are closed in $\mathbf{G}_n(C)$, since the latter is endowed with the topology induced from $\mathbf{M}_n(C)$. In view of Proposition 10, the proof is complete.

Theorem 11. $\mathbf{O}_n(R)$ *and* $\mathbf{U}_n$ *are compact topological subgroups of* $\mathbf{G}_n(C)$.

PROOF. Since $\mathbf{O}_n(R) \subset \mathbf{U}_n$ (Proposition 10) and since both are closed subsets of $\mathbf{G}_n(C)$ (Proposition 11), to prove the theorem it is sufficient to show that $\mathbf{U}_n$ is compact.

Let $A \in \mathbf{U}_n$. Then $\bar{A} = A^* = A'^{-1}$ or $A'\bar{A} = I$ (the identity matrix). This means that

$$\sum_{k=1}^{n} a_{ki} \bar{a}_{kj} = \delta_{ij},$$

where $\delta_{ij} = 1$ or 0 according as $i = j$ or $i \neq j$. For $i = j$, we have

$$\sum_{k=1}^{n} |a_{ki}|^2 = 1.$$

Therefore,

$$|a_{ki}| \leq 1$$

for all $i, k \leq n$. This shows that $\mathbf{U}_n$ is mapped onto a subset of I^{2n^2} under the identification mapping $f: \mathbf{M}_n(C) \to C^{n^2} = R^{2n^2}$, where $I = [0, 1]$. Since I^{2n^2} is bounded, so is $f(\mathbf{U}_n)$. But f being a homeomorphism and $\mathbf{U}_n$ being a closed subset of $\mathbf{M}_n(K)$ (proof of Proposition 11), it follows that $f(\mathbf{U}_n)$ is also closed in R^{2n^2}. But every closed and bounded set in R^{2n^2} is compact and so is $\mathbf{U}_n$ in $\mathbf{M}_n(C)$. But then the topology on $\mathbf{G}_n(C)$ being the one induced from $\mathbf{M}_n(C)$ and $\mathbf{U}_n \subset \mathbf{G}_n(C)$, the theorem is proved.

Notation 4. Let $\mathbf{SG}_n(C)$, $\mathbf{SO}_n(R)$ and $\mathbf{SU}_n$ denote the subsets of $\mathbf{G}_n(C)$, $\mathbf{O}_n(R)$, and $\mathbf{U}_n$, respectively, consisting of those matrices A for which $|A| = 1$. $\mathbf{SG}_n(R)$, $\mathbf{SO}_n(R)$ and $\mathbf{SU}_n$ *are called* special full linear, special orthogonal *and* special unitary groups, *respectively*.

The fact that the matrices whose determinants equal 1 form a group is easily verified.

Corollary 5. $\mathbf{SO}_n(R)$ *and* $\mathbf{SU}_n$ *are compact subgroups of* $\mathbf{G}_n(C)$.

PROOF. Since the mapping: $A \to |A|$ of $\mathbf{M}_n(C)$ into C is continuous (see proof of Proposition 9), $\mathbf{SO}_n(R)$ and $\mathbf{SU}_n$ are closed subsets of $\mathbf{M}_n(C)$, because they are inverse images of 1. Now from Theorem 11, the corollary follows.

28. LOCALLY EUCLIDEAN GROUPS

Our object in this section is to show that all classical linear groups described in §27 are locally Euclidean.

Definition 3.   A topological space E is said to be *locally Euclidean* if there exists a positive integer n such that every $x \in E$ has a neighborhood U homeomorphic to R^n.

Since R^n is homeomorphic with the unit open ball: $\left\{ x = (x_1, \ldots, x_n) \in R^n : \sum_{i=1}^{n} |x_i|^2 < 1 \right\}$, one can replace R^n by its unit open ball in the above definition.

Clearly a topological group is locally Euclidean if, and only if, the identity e lies in an open neighborhood U homeomorphic to R^n for some n.

Proposition 12. $\mathbf{M}_n(K)$ (§27) *is a locally Euclidean group.*

PROOF.   Since $K = R$ or C, $\mathbf{M}_n(K)$ is homeomorphic to R^{n^2} or $C^{n^2} = R^{2n^2}$ by the identification and the definition of the topology (see the proof of Proposition 8, §27). This completes the proof.

Now to show that other classical groups are locally Euclidean, we first discuss the exponential mapping *exp* defined as follows:
For $A = (a_{ij}) \in \mathbf{M}_n(K)$, let

$$\exp(A) = \sum_{m=0}^{\infty} \frac{A^m}{m!}, \qquad (m! = m(m-1)\cdots 1, \text{factorial } m),$$

where $A^0 = I$, the identity matrix and $A^m = AA \cdots A$ (m times).   Let $\sigma = \max_{1 \le i, j \le n} |a_{ij}|$. Then by induction one shows that the absolute value of each coefficient of A^m is less than or equal to $(n\sigma)^m$. Hence, $\sum_{m=0}^{\infty} \frac{a_{ij}^{(m)}}{m!}$ is convergent for each i, j $(1 \le i, j \le n)$, where $a_{ij}^{(m)}$ is the (i,j)th coefficient of A^m. This shows that $\sum_{m=0}^{\infty} \frac{A^m}{m!}$ is always convergent and, hence, $\exp(A)$ is well-defined. Actually, $\sum_{m=0}^{\infty} \frac{A^m}{m!}$ converges uniformly on any compact subset of $\mathbf{M}_n(K)$. Thus it follows that $\exp(A)$ is a continuous mapping of $\mathbf{M}_n(K)$ into itself.

A few properties of $\exp(A)$ are given in the following:

Proposition 13.   (a) *For each* $B \in \mathbf{G}_n(K)$, $\exp(BAB^{-1}) = B\exp(A)B^{-1}$.
(b) *If* $AB = BA$, *then* $\exp(A + B) = \exp(A)\exp(B)$.
(c) $\exp(A') = (\exp(A))'$.
(d) $\exp(\bar{A}) = \overline{\exp(A)}$.

(e) *If λ_i $(1 \leq i \leq n)$ are characteristic roots of A, then $\exp \lambda_i$ $(1 \leq i \leq n)$ are characteristic roots of $\exp (A)$.*

(f) $|\exp (A)| = \exp (T_r A)$, *where* $T_r A = $ *Trace of* $A = \sum_{i=1}^{n} a_{ii}$.

(g) *For every* $A \in \mathbf{M}_n(K), \exp (A) \in \mathbf{G}_n(K)$. *And* $\exp (-A) = [\exp (A)]^{-1}$.

PROOF. (a) Since B is regular, it is easy to see that $BA^m B^{-1} = (BAB^{-1})^m$. Therefore,

$$\exp (BAB^{-1}) = \sum_{m=0}^{\infty} \frac{(BAB^{-1})^m}{m!} = \sum_{m=0}^{\infty} B\left(\frac{A^m}{m!}\right)B^{-1} = B \exp (A)B^{-1}.$$

(b) If $AB = BA$, then $(A + B)^m$ may be written:

$$(A + B)^m = \sum_{k=0}^{m} \frac{m!}{k!(m - k)!} A^k B^{m-k}.$$

Hence, for any integer p, we have

$$\sum_{m=0}^{2p} \frac{1}{m!} (A + B)^m = \sum_{m=0}^{2p} \left(\sum_{k=0}^{m} \frac{A^k}{k!} \frac{B^{m-k}}{(m - k)!} \right)$$

$$= \left(\sum_{k=0}^{p} \frac{A^k}{k!} \right)\left(\sum_{k=0}^{p} \frac{B^k}{k!} \right) + R_p,$$

where R_p is the sum $\sum_{(k,j)} \frac{A^k}{k!} \frac{B^j}{j!}$ taken over all combinations (k, j) such that $\max (k, j) > p$ and $k + j \leq 2p$. The number of such combinations is $p(p + 1)$. Moreover, if $\sigma = \max_{i,j} (|a_{ij}|, |b_{ij}|)$, $A = (A_{ij})$, $B = (b_{ij})$, then the absolute value of any coefficient of $\frac{A^k B^j}{k! j!}$ is less than or equal to

$$\frac{n(n\sigma)^k(n\sigma)^j}{k!j!} = \frac{n(n\sigma)^{k+j}}{k!j!} \leq \frac{(n\sigma_0)^{2p}}{p!},$$

where σ_0 is a positive number independent of p. Hence, the absolute value of R_p is bounded by $\dfrac{p(p + 1)(n\sigma_0)^{2p}}{p!}$ which approaches zero as $p \to \infty$. Whence, we have

$$\sum_{m=0}^{\infty} \frac{1}{m!} (A + B)^m = \left(\sum_{k=0}^{\infty} \frac{A^k}{k!} \right)\left(\sum_{k=0}^{\infty} \frac{B^k}{k!} \right)$$

This proves (b).

(c) $\exp (A') = \sum_{m=0}^{\infty} \frac{A'^m}{m!} = \sum_{m=0}^{\infty} \frac{(A^m)'}{m!} = \left(\sum_{m=0}^{\infty} \frac{A^m}{m!} \right)' = (\exp (A))'.$

(d) Similar to (c).

(e) The proof is by induction on the order n of the matrix A. If $n = 1$, A is a complex number and, hence, the characteristic root is this complex number itself, whence (e) is trivial. Suppose (e) is true for all matrices of order $n - 1$. Let A be any square matrix of order n and let λ_1 be one of its characteristic roots. Then from linear algebra it is known that there exists a regular matrix $B \in \mathbf{G}_n(K)$ such that BAB^{-1} is a matrix whose first column is $(\lambda_1, 0, \ldots, 0)$. But then it follows that the first column of $\frac{1}{m!}(BAB^{-1})^m$ is $\left(\frac{\lambda_1^m}{m!}, 0, \ldots, 0\right)$, for any nonnegative integer m. Hence, the first column of $\exp(BAB^{-1}) = B\exp(A)B^{-1}$, (part (a)), is $(e^{\lambda_1}, \ldots, 0)$. Since the matrix obtained by deleting the first row and first column of $B\exp(A)B^{-1}$ is of order $(n-1)$ and hence has characteristic roots $e^{\lambda_2}, \ldots, e^{\lambda_n}$ by induction, and since the characteristic roots of $\exp(A)$ and $B\exp(A)B^{-1}$ are the same, (e) is established.

(f) For any matrix $A \in \mathbf{M}_n(K)$, $|A| = \prod\limits_{i=1}^{n} \lambda_i$, where λ_i $(1 \leq i \leq n)$ are characteristic roots of A, and $T_rA = \sum\limits_{i=1}^{n} \lambda_i$, as is well known from linear algebra. Hence, $|\exp(A)| = \prod\limits_{i=1}^{n} \exp \lambda_i = \exp\left(\sum\limits_{i=1}^{n} \lambda_i\right) = \exp(T_rA)$.

(g) For any $A \in \mathbf{M}_n(K)$, by (f) $|\exp(A)| = \exp(T_r(A)) \neq 0$, and so $\exp(A) \in \mathbf{G}_n(K)$. Furthermore by (b),

$$\exp(A) \times \exp(-A) = \exp(A - A) = \exp(0) = I,$$

where 0 is the zero matrix and I the identity matrix. This completes the proof of (g).

Theorem 12. $\mathbf{G}_n(K)$, and all its subgroups mentioned in §27 are locally Euclidean.

PROOF. Since $\mathbf{G}_n(K)$ is an open subset of $\mathbf{M}_n(K)$ and since $\mathbf{M}_n(K)$ is locally Euclidean (Proposition 12), it is clear that $\mathbf{G}_n(K)$ is also locally Euclidean. To show that the above-mentioned subgroups of $\mathbf{G}_n(K)$ are locally Euclidean, we need the following: there exists an open neighborhood U of 0 in $\mathbf{M}_n(K)$ such that U is homeomorphic to a neighborhood of I in $\mathbf{G}_n(K)$ under the mapping exp. For each $A \in \mathbf{M}_n(K)$, $\exp(A) \in \mathbf{G}_n(K)$ by Proposition 13(g). Thus, under a fixed arrangement of coefficients of matrices in $\mathbf{M}_n(K)$, $A = (a_{ij})$ $(1 \leq i,j \leq n)$ is mapped onto $\exp(A) = (f_{ij})$, $(1 \leq i,j \leq n)$, where f_{ij} is an entire function of a_{ij}'s. Thus, exp can be regarded as a mapping of K^{n^2} into K^{n^2}. Since $\exp(A) = \sum\limits_{m=0}^{\infty} \frac{A^m}{m!}$, clearly

$$f_{ij}(a_1, \ldots, a_{nn}) = \delta_{ij} + a_{ij} + \sum$$

where $\sum$ is the summation over terms of higher powers of $a_{11}, \ldots, a_{nn}$ in the Maclaurin expansion of f_{ij}. The Jacobian of exp at 0 is equal to 1, and therefore, by the implicit function theorem, there exists an open neighborhood U of 0 in $\mathbf{M}_n(K)$ and that $\exp(U)$ is an open neighborhood V of I in $\mathbf{G}_n(K)$ and where the mapping exp is a homeomorphism.

Among subgroups of $\mathbf{G}_n(K)$, we choose $\mathbf{U}_n$ and show that it is locally Euclidean. The remaining ones can be treated similarly (e.g., see Chevalley[9]).

Let U be a neighborhood of 0 in $\mathbf{M}_n(K)$ such that exp is a homeomorphism of U (see the above paragraph) onto a neighborhood of I in $\mathbf{G}_n(K)$. Since the mappings: $A \to A'$, $A \to \bar{A}$, and $A \to -A$ are continuous in $\mathbf{M}_n(K)$, we may assume that for each A in U, A', $\bar{A}$, and $-A$ are also in U. Let $W = \{A \in U : A' = -\bar{A}\}$ which is the set of all skew-Hermitian matrices in U. It is known that a matrix $A = (a_{ij})$ is skew-Hermitian if, and only if, a_{ii} is purely imaginary and $a_{ij} = -\bar{a}_{ji}$. Thus, the subset **SH** of all skew-Hermitian matrices in $\mathbf{M}_n(K)$ is a real Euclidean space of dimension n^2. Therefore, W is an open neighborhood of 0 in R^{n^2}. Now we show that exp maps W onto a neighborhood of I in $\mathbf{U}_n$ homeomorphically. For this let $A \in W$. Then

$$(\exp(A))^* = ((\exp(A))')^{-1} = (\exp(A'))^{-1} = \exp(-A')$$
$$= \exp(\bar{A}) = \overline{\exp(A)},$$

by Proposition 13. Hence, $\exp(A) \in \mathbf{U}_n$. On the other hand, if $\exp(A) \in \mathbf{U}_n \cap \exp U$, then $-A' = \bar{A}$, or $A \in \mathbf{SH}$. Hence, exp maps W onto a neighborhood of I in $\mathbf{U}_n$.

The dimensions of Euclidean spaces that are homeomorphic to neighborhoods of the identity in $\mathbf{G}_n(C)$, $\mathbf{S}_n(C)$, $\mathbf{U}_n$, $\mathbf{G}_n(R)$, $\mathbf{S}_n(R)$, $\mathbf{O}_n(C)$, and $\mathbf{O}_n$ are

$2n^2$, $2n^2 - 2$, n^2, n^2, $n^2 - 1$, $n(n-1)$, and $\dfrac{n(n-1)}{2}$, respectively.

There exist topological groups that are not locally Euclidean. For example, let $G = \prod\limits_\alpha R_\alpha$, the direct product of an infinite number of copies R_α of the real line R, endowed with the product topology. Then G is clearly not locally Euclidean.

29. LIE GROUPS

Definition 4. (a) Let f be a real-valued function defined on an open subset S of R^n. Then f is said to be in *the class ∞ or C^∞*, if all mixed partial derivatives of all orders exist and are continuous on S.

(b) Let X be a topological space. An *atlas* $\mathscr{A}$ of class C^∞ on X is a collection of pairs (U_α, ψ_α), $\alpha \in A$, an index set, satisfying the following conditions:

(i) Each U_α is an open subset of X and $\{U_\alpha\}$ covers X.

(ii) Each ψ_α is a homeomorphism of U_α on an open subset $\psi_\alpha(U_\alpha)$ of R^n for some fixed n.

(iii) For $U_\alpha \cap U_\beta \neq \varnothing$, $\psi_\alpha \circ \psi_\beta^{-1} : \psi_\beta(U_\alpha \cap U_\beta) \to \psi_\alpha(U_\alpha \cap U_\beta)$ is of class C^∞.

(iv) Let (U, ψ) be a pair consisting of an open subset U of X and a homeomorphism ψ of U onto an open subset of R^n. If for each pair $(U_\alpha, \psi_\alpha) \in \mathscr{A}$ for which $U_\alpha \cap U \neq \varnothing$, the map $\psi \circ \psi_\alpha^{-1} : \psi_\alpha(U \cap U_\alpha) \to \psi(U \cap U_\alpha)$ is of C^∞, then $(U, \psi) \in \mathscr{A}$.

(c) A Hausdorff topological space X with an atlas $\mathscr{A}$ is called an *n-dimensional C^∞ manifold* or simply a *manifold*. The members of the atlas are usually called the *coordinate system* for the manifold.

The following is immediate from the above definitions.

Proposition 14. *Every manifold is locally Euclidean and therefore locally compact.*

The converse problem—when does a locally Euclidean space become a manifold?—runs deeper. For topological groups, as we shall state in the sequel, the problem in the modified form is known as the famous fifth problem of Hilbert. Apart from mentioning the latter problem, we shall not make any attempt to present its solution. The reader is referred to Montgomery and Zippin,[34,35] and Gleason[14] for further details.

A finite product of manifolds is also a manifold. For a proof, see Chevalley,[9] Chapter III.

Let P be an open set in the Euclidean space R^m. A mapping $f: P \to R^n$ is said to be in *the class* ∞ *or* C^∞ if, for each C^∞ real-valued function φ defined on an open set in $f(P)$, $\varphi \circ f$ is of C^∞ in the sense of Definition 4.

Let M and N be two manifolds of dimensions m and n, respectively. A mapping $f: M \to N$ is said to be of C^∞ if there exists an atlas $\{U_\alpha, \varphi_\alpha\}$ of M and an atlas $\{V_\beta, \varphi_\beta\}$ of N such that for each α, β the mapping

$$\psi_\beta \circ f \circ \varphi_\alpha^{-1}: \varphi_\alpha(U_\alpha) \to \psi_\beta(V_\beta)$$

is of C^∞ in the sense of the previous paragraph. (Observe that $\varphi_\alpha(U_\alpha)$ and $\psi_\beta(V_\beta)$ are open subsets of R^m and R^n, respectively.)

If the mapping f in Definition 4 has a power series expansion, then f is said to be *analytic*.

Thus, one can define analytic manifolds and analytic mappings between manifolds, etc., just by replacing C^∞ mappings involved in the definitions by analytic mappings.

Definition 5. A manifold G that is also a group is called a *Lie group* if the mappings:
  (i) $(x, y) \to xy$ of $G \times G$ onto G
  (ii) $x \to x^{-1}$ of G onto G
are analytic.

Examples. 1. R^n is an additive abelian Lie group. For each $x \in R^n$, $x = (x_1, \ldots, x_n)$, put $\psi(x) = (x_1(x), \ldots, x_n(x)) \in R^n$, where $x_i = x_i(x)$ are real functions of x. Then we see that ψ is in C^∞ and all the conditions in the definition of a manifold are satisfied. Hence, R^n is a manifold. Indeed, it is an additive abelian group. Furthermore, the mappings: $(x, y) \to x + y$ and $x \to -x$ are analytic. Hence, R^n is a Lie group.

2. Since $\mathbf{M}_n(K)$ can be identified with R^{n^2} or R^{2n^2} according as K is R or C, $\mathbf{M}_n(K)$ is also a manifold and, hence, a Lie group.

Now by inducing the structure of the manifold from $\mathbf{M}_n(K)$ on $\mathbf{G}_n(K)$, $\mathbf{U}_n$ etc., one sees that they are Lie groups as well.

One can prove the following:

Proposition 15.   (a) *A closed subgroup of a Lie group is a Lie group.*

(b) *Let G be a Lie group and H a closed invariant subgroup of G. Then G/H is a Lie group.*

(c) *A finite product of Lie groups is a Lie group.*

For the proof, the reader is referred to Chevalley,[9] Chapter IV.

In view of Proposition 14, every Lie group is locally Euclidean and, hence, locally compact. The well-known fifth problem of Hilbert states that every locally Euclidean topological group is a Lie group. For compact and abelian topological groups the problem was solved long before the general solution was found (see Pontrjagin[37]). For details of the complete solution of this problem, the reader is referred to Montgomery and Zippin,[34,35] and Gleason.[14]

Exercises

1. Let G be a locally compact totally disconnected topological group and H a closed subgroup of G. The space G/H of cosets is totally disconnected.

2. Let G be a topological group and H a subgroup of G. If H and G/H are locally compact, so is G.

3. Let G be a locally compact group and H a Hausdorff topological group. Let f be a continuous and open homomorphism of G onto H. Let M be the component of the identity in G. Then $\overline{f(M)}$ is the component of the identity in H.

4. Prove Propositions 14 and 15.

5. (Structure theorems.) Let G be a locally compact Hausdorff abelian group such that for some compact symmetric neighborhood U of its identity e, $G = \bigcup\limits_{n=1}^{\infty} U^n$. Then:

(a) G contains a discrete subgroup N with a finite number of linearly independent generators such that G/N is compact.

(b) Each neighborhood of e contains a compact subgroup H such that G/H is homeomorphic with $T^n \times R^m \times Z^l \times F_0$, where $n, m, l \geq 0$, integers and F_0 a finite abelian group.

(c) G is homeomorphic with $R^{n_1} \times Z^{n_2} \times F$, for some $n_1, n_2 \geq 0$ integers and some compact abelian group F.

V

Open Homomorphisms and Closed Graphs

Let E and F be two semitopological or topological groups. The following statements may or may not be true:

(o.h.) A continuous homomorphism f of E onto F is open.

(c.g.) A homomorphism g of E into F, the graph of which is closed in $E \times F$, is continuous.

Whenever the statements (o.h.) and (c.g.) are true for some pair E and F with or without additional conditions on f and g, they are said to be the *open homomorphism* and *closed graph* theorems, respectively.

In this chapter, we shall establish (o.h.) and (c.g.) for a large class of topological groups.

30. CONTINUOUS AND OPEN HOMOMORPHISMS

The following lemmas are useful in the sequel.

Lemma 1. *Let E and F be two groups and f a homomorphism of E into F. Then:*

(a) *for any subsets A and B of E, $f(AB) = f(A)f(B)$;*

(b) *for any subsets C and D of F, $f^{-1}(C)f^{-1}(D) \subset f^{-1}(CD)$;*

(c) *for any symmetric subset A of E, $f(A)$ is symmetric in F;*

(d) *for any symmetric subset B of F, $f^{-1}(B)$ is symmetric in E.*

PROOF.   (a) is obvious because f is a homomorphism.

(b) Let $x \in f^{-1}(C)$ and $y \in f^{-1}(D)$. Then $f(x) \in C$ and $f(y) \in D$. Hence, $f(xy) = f(x)f(y) \in CD$ shows that $xy \in f^{-1}(CD)$, since f is a homomorphism.

(c) We have to show that $f(A) = [f(A)]^{-1}$. Let $y \in f(A)$. Then $y = f(x)$, $x \in A$. Symmetry of A implies $x^{-1} \in A$ and, hence, $y^{-1} = [f(x)]^{-1} = f(x^{-1}) \in f(A)$. Or $y \in [f(A)]^{-1}$. This shows $f(A) \subset [f(A)]^{-1}$. For the reverse inclusion, let $x \in [f(A)]^{-1}$. Then $x^{-1} \in f(A)$ and so $x^{-1} = f(y)$, $y \in A$. Again using the symmetry of A, we have $y^{-1} \in A$ and, therefore, $x = [f(y)]^{-1} = f(y^{-1}) \in f(A)$. Thus (c) is established.

(d) The proof is similar to that of (c).

Lemma 2.   *Let E and F be two topological groups and f a homomorphism of E into F. Then*

(a') *for any subsets A and B of E, $\overline{f(A)}\,\overline{f(B)} \subset \overline{f(AB)}$*

(b') *for any subsets C and D of F, $\overline{f^{-1}(C)}\,\overline{f^{-1}(D)} \subset \overline{f^{-1}(CD)}$*

(c') *for any symmetric subset A of E, $\overline{f(A)}$ is symmetric in F and hence* $\overline{f(A^{-1})} = [\overline{f(A)}]^{-1}$

(d') *for any symmetric subset B of F, $\overline{f^{-1}(B)}$ is symmetric in E and hence* $\overline{f^{-1}(B)} = [\overline{f^{-1}(B)}]^{-1}.$

PROOF.   (a') and (b') follow from (a) and (b) of Lemma 1, respectively, and the fact that in a topological group, for any subsets A and B, $\bar{A}\bar{B} \subset \overline{AB}$, since the multiplication is continuous.

For (c'), observe that $f(A)$ is symmetric ((c) of Lemma 1). Moreover, the inversion mapping in a topological group being a homeomorphism, for each subset A, we have $\overline{A^{-1}} = (\bar{A})^{-1}$. This proves (c'), and (d') follows similarly.

The following proposition shows that the continuity and openness of homomorphisms at the identity imply the continuity and openness of homomorphism at all points.

Proposition 1.   *Let $\{U\}$ and $\{V\}$ denote fundamental systems of neighborhoods of e and e' in semitopological groups E and F, respectively. Let f be a homomorphism of E into F. Then*

(a) *f is continuous if, and only if, for each V in $\{V\}$, $f^{-1}(V)$ is a neighborhood of e in E.*

(b) *f is open if, and only if, for each U in $\{U\}$, $f(U)$ is a neighborhood of e' in F.*

PROOF.   (a) If f is continuous then the condition is obvious. Conversely, let P be an open set of F such that $y = f(x) \in P$, $x \in E$. Then $y^{-1}P$

is a neighborhood of e' and therefore there exists a V in $\{V\}$ such that $V \subset y^{-1}P$. But then, by assumption, there exists a U in $\{U\}$ such that $U \subset f^{-1}(V)$ or $f(U) \subset V \subset y^{-1}P$. Since xU is a neighborhood of x (Theorem 1, §12, Chapter II) and $f(xU) = f(x)f(U) \subset yy^{-1}P = P$, this proves the continuity of f.

(b) The proof is similar to that given for (a).

Proposition 2. *Let f be a homomorphism of a Hausdorff semitopological group E into another Hausdorff semitopological group F such that the graph of f is closed in $E \times F$. Let $H = \{x \in E : f(x) = e'\}$. Then H is a closed invariant subgroup of E.*

PROOF.   It is well known that H is an invariant subgroup of E. Let $x \in \bar{H}$, and let G denote the graph of f in $E \times F$. Let U and V be any neighborhoods of e and e' in E and F, respectively. Then clearly there exists $x_1 \in H$ such that $x_1 \in xU$ and $f(x_1) \in V$. In other words, $(x_1, f(x_1)) \in (xU, V) \cap G \neq \varnothing$, i.e., $(x, e') \in \bar{G}$. Since G is closed by hypothesis, $(x, e') \in G$. This shows that $f(x) = e'$, or $x \in H$.

Proposition 3. *Let f be a homomorphism of a semitopological group E into another semitopological group F. Let $H = f^{-1}(e')$, where e' is the identity of F. Let $\varphi : E \to E/H$ be the canonical map and $f = f_0 \circ \varphi$ for some $f_0 : E/H \to F$. Then:*

(a) *f is continuous if, and only if, f_0 is;*
(b) *f_0 is almost continuous if f is;*
(c) *f is open if, and only if, f_0 is;*
(d) *f is almost open if, and only if, f_0 is;*
(e) *f_0 is a $1:1$ homomorphism of E/H into F.*

PROOF.   By the definition of the quotient topology on E/H and by Theorem 8, §24, Chapter III, it follows that φ is a continuous and open homomorphism of E onto the semitopological group E/H, since H is an invariant subgroup and G is a semitopological group.

(a) If f_0 is continuous then f, being the composition of two continuous homomorphisms, is continuous.

Conversely, if f is continuous, then for each neighborhood V of e' in F, $f_0^{-1}(V) = \varphi(f^{-1}(V))$ is a neighborhood of e in E, because φ is open. Hence, f_0 is continuous.

(b) Similarly, as in (a), for each neighborhood V of e' in F,

$$f_0^{-1}(V) = \varphi(f^{-1}(V)),$$

and so

$$\overline{f_0^{-1}(V)} = \overline{\varphi(f^{-1}(V))} \supset \varphi(\overline{f^{-1}(V)}),$$

because of the continuity of φ. Now if f is almost continuous, $\overline{f^{-1}(V)}$ is a neighborhood of e in E, and hence $\varphi(\overline{f^{-1}(V)})$ is a neighborhood of $\dot{e}$ in

E/H because φ is open. Hence, $\overline{f_0^{-1}(V)}$ is a neighborhood of $\dot{e}$. This proves that f_0 is almost continuous.

(c) For each open neighborhood U of e in E,

$$f(U) = f_0(\varphi(U)),$$

and therefore (c) is established, because φ is open.

(d) From the equation in (c), we have

$$\overline{f(U)} = \overline{f_0(\varphi(U))}.$$

Since φ is open it follows that f_0 is almost open if, and only if, f is so.

(e) The proof is simple and is left to the reader.

Theorem 1. *Let E and F be two semitopological groups and f a continuous and open homomorphism of E onto F. Let $H = f^{-1}(e')$, where e' is the identity of F. Then F is homeomorphically isomorphic to E/H.*

PROOF. As pointed out earlier, H is an invariant subgroup of E and so E/H is a quotient group. E/H is a semitopological group since E is so. Also E/H and F are isomorphic, say via f_0 in the above proposition, as is known from group theory. By Proposition 3, $f_0 : E/H \to F$ is continuous, open, $1:1$ and onto.

Proposition 4. *Let f be a continuous homomorphism of a topological group E into another topological group F. Then f is uniformly continuous whenever E and F are regarded as the uniform spaces having either both left (or both right) uniform structures.*

PROOF. Suppose f is continuous at $x_0 \in E$. Let V be a neighborhood of e' in F. Then by the continuity of f there exists a neighborhood U of e in E such that $f(x_0 U) \subset f(x_0)V$. Thus, if $y^{-1}x \in U$, we have

$$f(x_0)[f(y)]^{-1}f(x) = f(x_0 y^{-1}x) \in f(x_0 U) \subset f(x_0)V.$$

Or $[f(y)]^{-1}f(x) \in V$. This proves that f is uniformly continuous when E and F both are endowed with the left uniform structures.

One asks the following question: What can be said about the uniform continuity of a real-valued continuous function on a topological group? Indeed, in general, a real valued continuous function on the real line is not necessarily uniformly continuous. For example, $f(x) = x^2$. However, the following is true:

Proposition 5. *Let E be a locally compact Hausdorff topological group and f a real-valued continuous function on E such that $C = \overline{\{x \in E : f(x) \neq 0\}}$ is a compact subset of E. Then f is left and right uniformly continuous.*

PROOF. Let U be a compact symmetric neighborhood of e in E. Such a neighborhood exists because E is locally compact. Let $\varepsilon > 0$ be given.

Consider the set $W = \{y \in E : |f(yx) - f(x)| < \varepsilon$, for $x \in UC\}$. First we show that W is a neighborhood of $e \in E$. Clearly, $e \in W$. Since f is continuous, there exist open neighborhoods U_x of $x \in UC$ and V_x of e such that for all $y \in V_x$ and $z \in U_x$,

$$|f(yz) - f(x)| < \varepsilon.$$

As x runs over UC, $\{U_x\}$ forms an open covering of UC. Owing to compactness of UC (both U and C are compact), the union of a finite subfamily $\{U_{x_i}\}$ $(1 \leq i \leq n)$ covers UC. Let $\{V_{x_i}\}(1 \leq i \leq n)$ be the V's corresponding to $\{U_{x_i}\}$ $(1 \leq i \leq n)$. Then $V = \bigcap_{i=1}^{n} V_{x_i}$ is an open neighborhood of e, and for each $y \in V$ and $x \in UC$,

$$|f(yx) - f(x)| < \varepsilon.$$

But then by the definition of W, it follows that $e \in V \subset W$. Hence, W is a neighborhood of e.

Clearly, $f(x) = 0$ for all $x \in E \sim UC$. Moreover, if $y \in U$ and $x \in E \sim UC$, then $f(yx) = 0$ because, otherwise, $yx \in C$ or $x \in y^{-1}C \subset UC$, which is impossible. Hence, for all $y \in U \cap W$ and for all $x \in E$, we have

$$|f(yx) - f(x)| < \varepsilon.$$

Since $U \cap W$ is a neighborhood of e, this proves that f is left uniformly continuous.

Similarly one proves right uniform continuity.

31. B(𝒞) GROUPS

In this section, we introduce the notion of $B(\mathscr{C})$ groups—a notion introduced by the author[20] for locally convex linear spaces.

It turns out that the closed graph and the open homomorphism theorems proved for $B(\mathscr{C})$ linear spaces do not really need extra algebraic structures provided by the scalar multiplication. This section demonstrates that the concept of $B(\mathscr{C})$ spaces can be carried over to groups, and thus one can prove the open homomorphism and closed graph theorems for $B(\mathscr{C})$ groups as well.

Definition 1. Let $\mathscr{C}$ be a class of Hausdorff topological groups. A Hausdorff topological group E is said to be a $B(\mathscr{C})$ *group*, if for each $F \in \mathscr{C}$, a continuous and almost open homomorphism f of E onto F is open.

If f, in addition, is $1:1$ in the above definition, then E is said to be a $B_r(\mathscr{C})$ group.

See author's papers[20,22] for these notions in linear spaces.

The following is immediate:

Proposition 6. *If $\mathscr{C}_1$ and $\mathscr{C}_1$ are two classes of Hausdorff topological groups such that $\mathscr{C}_1 \subset \mathscr{C}_2$, then every $B(\mathscr{C}_2)$ group is also a $B(\mathscr{C}_1)$ group.*

This proposition implies that the class of all $B(\mathscr{A})$ groups is the smallest of all classes of $B(\mathscr{C})$ groups, where $\mathscr{A}$ is the class of *all* Hausdorff topological groups.

V. Pták[38] called $B(\mathscr{A})$ locally convex linear spaces B-complete locally convex spaces. The term B-completeness for locally convex spaces is justified, since each $B(\mathscr{A})$ locally convex linear space is complete. However, I do not know if this is true for $B(\mathscr{A})$ groups as well. The following two theorems indicate that a good number of topological groups are $B(\mathscr{A})$ groups.

Theorem 2. *Every compact Hausdorff semitopological group E is a $B(\mathscr{A})$ group and, therefore, a $B(\mathscr{C})$ group for any class $\mathscr{C}$ of Hausdorff topological groups.*

PROOF. By Exercise B (1), Chapter II, E is a topological group. Let F be any Hausdorff topological group and f a continuous almost open homomorphism of E onto F. Since f is continuous, $f^{-1}(e') = H$ (where e' is the identity of F) is a closed invariant subgroup. Since E/H is compact, for our purpose it is sufficient to show that the mapping: $E/H \to F$ is open. But this follows from Proposition 3, §30, because $E/H \to F$, obviously a continuous $1:1$ mapping of a compact space onto a Hausdorff topological group, is a homeomorphism.

The following is a particular case of the above theorem.

Corollary 1. *Every compact metric semitopological group is a $B(\mathscr{A})$ group.*

The fact that even a Hausdorff locally compact topological group is a $B(\mathscr{A})$ group is left as an exercise for the reader.

Theorem 3. *Every metrizable complete topological group E is a $B(\mathscr{A})$ group and, hence, a $B(\mathscr{C})$ group for any class $\mathscr{C}$ of Hausdorff topological groups.*

PROOF. Let $\{U_n\}$ be a countable fundamental system of closed symmetric neighborhoods of e in E such that;

(i) $U_{n+1}^2 \subset U_n$ for each $n \geq 1$;

(ii) $\bigcap_{n \geq 1} U_n = \{e\}$.

Let F be any Hausdorff topological group and f a continuous almost open homomorphism of E onto F. Then for each $n \geq 1$, $W_n = \overline{f(U_n)}$ is a neighborhood of $e' \in F$, because f is almost open. Since $U_{n+1} \subset U_n$, $W_{n+1} \subset$

W_n for each n. Now first we show that

(I)
$$\bigcap_{n \geq 1} W_n = \{e'\}.$$

Let $y \in W_n$ for each n. Then for any closed symmetric neighborhood V of e' in F, there exists $x_n \in U_n$ such that $f(x_n) \in yV$. Since U_n is a decreasing sequence of a fundamental system of neighborhoods of e in E, $x_n \to e$. Since f is continuous, $f(x_n) \to f(e) = e'$. Since V is closed, $e' \in yV$. The symmetry of V implies that $y \in V$. Since V is arbitrary and F Hausdorff, it follows that $y = e'$.

Now to prove the theorem it is sufficient to show that $f(U_n)$ is a neighborhood of e' in F for each $n \geq 1$. For this, it will suffice to show that $f(U_k) \supset W_{k+1}$ for each k.

Let $y \in W_{k+1}$. Since $f(U_{k+1})$ is dense in W_{k+1}, there exists y_1 such that $f^{-1}(y_1) = x_1 \in U_{k+1}$ and $y_1 \in yW_{k+2}$. Proceeding inductively, we produce for each $n \geq 1$ a y_n such that $f^{-1}(y_n) = x_n \in U_{k+n}$ and such that $y_1 y_2 \cdots y_n \in yW_{k+n+1}$. This means:

(II)
$$\prod_{i=1}^{n} y_i \in yW_{k+n+1}, \qquad \text{for all } n \geq 1.$$

Since

$$x_n x_{n+1} \cdots x_{n+p} \in U_{k+n} U_{k+n+1} \cdots U_{k+n+p}$$

and by (i)

$$U_{k+n} \supset U_{k+n+1} U_{k+n+2} \cdots U_{k+n+p},$$

we conclude

(III)
$$\prod_{i=0}^{p} x_{n+i} \in U_{k+n} U_{k+n} \subset U_{k+n-1}, \qquad \text{for all } p \geq 0.$$

Now if U is any arbitrary neighborhood of e in E, there exists n_0 such that for all $n \geq n_0$ and $p \geq 0$, $\prod_{j=n}^{n+p} x_j = \prod_{i=0}^{p} x_{n+i} \in U$. This shows that $\{S_n\} = \left\{\prod_{i=1}^{n} x_i\right\}$ forms a Cauchy sequence. Since E is complete, there exists $x \in E$ such that $x = \lim_{n \to \infty} S_n = \prod_{i=1}^{\infty} x_i$.

From (III) it follows that (by putting $n = 1$)

$$\prod_{i=1}^{p} x_i \in U_k, \qquad \text{for all } p \geq 1.$$

Since U_k is closed, $x = \prod_{i=1}^{\infty} x_i \in U_k$. But since f is a continuous homomorphism,

$$f(x) = \prod_{i=1}^{\infty} f(x_i) = \prod_{i=1}^{\infty} y_i$$

because $f(x_i) = y_i$ for all $i \geq 1$.

Also from (II) it follows that for large m,

$$y^{-1}\left(\prod_{i=1}^{m} y_i\right) \in W_{k+n+1},$$

since $\{W_n\}$ is a decreasing sequence. But W_{k+n+1} being closed,

$$y^{-1}\left(\prod_{i=1}^{\infty} y_i\right) \in W_{k+n+1}$$

for each $n \geq 1$. Hence, using (I), we have

$$y^{-1}\left(\prod_{i=1}^{\infty} y_i\right) \in \bigcap_{n \geq 1} W_{k+n+1} = \{e'\},$$

or

$$y = \prod_{i=1}^{\infty} y_i = \prod_{i=1}^{\infty} f(x_i) = f(x).$$

But $x \in U_k$ implies $y = f(x) \in f(U_k)$. Since $y \in W_{k+1}$ by assumption, this shows that $f(U_k) \supset W_{k+1}$, and hence the theorem is proved.

Corollary 2. *Every complete metric separable semitopological group E is a $B(\mathscr{A})$ group.*

PROOF. By Corollary 4, §17, Chapter II, E is a topological group. Therefore by the above theorem, E is a $B(\mathscr{A})$ group.

Proposition 7. *Let E be a $B(\mathscr{C})$ group (or $B_r(\mathscr{C})$ group) and H any closed invariant subgroup of E. Then E/H is also a $B(\mathscr{C})$ group (or $B_r(\mathscr{C})$ group).*

PROOF. Let F be any Hausdorff topological group in $\mathscr{C}$ and f a continuous almost open homomorphism of E/H onto F. Let φ denote the canonical mapping: $E \to E/H$. Let $g = f \circ \varphi$. Then g is clearly a continuous homomorphism of E onto F, because both f and φ are continuous and onto. By Proposition 3, §30, g is almost open because f is. Therefore, g is open because E is a $B(\mathscr{C})$ group. Hence, f is open (Proposition 3, §30), and so E/H is a $B(\mathscr{C})$ group.

The same argument works for $B_r(\mathscr{C})$ groups.

Let u and v be two Hausdorff topologies on a group E such that E_u and E_v are semitopological (or topological) groups. Let $\{U\}$ denote a fundamental system of u-neighborhoods of e. Let $Cl_v U$ denote the v-closure of each U in $\{U\}$. As in §16, Chapter II, let $v(u)$ denote the topology on E for which $\{Cl_v U\}$ forms a fundamental system of neighborhoods of e in E. It is easy to see that $E_{v(u)}$ is a semitopological or topological group, according as E_u is a semitopological or topological group.

With the above notations we have:

Proposition 8. *Let u and v be two topologies on a set E such that E_u and E_v are semitopological groups and $u \supset v$. If E_v is a T_3-space, then $u \supset v(u) \supset v$.*

PROOF. Since for each $Cl_v U \supset U$, we conclude $u \supset v(u)$. Now, E_v

being a T_3-space, let $\{V\}$ denote a fundamental system of v-closed v-neighborhoods of e. For each V in $\{V\}$, there exists a U such that $U \subset V$ and hence $Cl_v U \subset V$, because V is v-closed. This shows that $v(u) \supset v$.

Proposition 9. *Let i be the continuous identity homomorphism of a semitopological group E_u onto a T_3 semitopological group E_v (i.e., $u \supset v$ on E). Then i is almost open if, and only if, $v(u) = v$.*

Proof.   Since $u \subset v$, by Proposition 8 $u \supset v(u) \supset v$. Now if i is almost open then for each U, $Cl_v U$ is a v-neighborhood of e, i.e., there exists a V such that $V \subset Cl_v U$. This shows that $v(u) \subset v$. Hence, $v(u) = v$. The "if" part is obvious.

Theorem 4.   *If E_u is a $B(\mathscr{A})$ group then for each Hausdorff topology v on E such that E_v is a topological group, $u \supset v$ and $v(u) = v$ together imply $u = v$.*

Proof.   Let $F = E_v$ and i be the identity homomorphism of E_u onto F. Then i is continuous because $u \supset v$, and almost open because $v(u) = v$ (Proposition 9) and because every Hausdorff topological group is a T_3-space. Thus i is open i.e., $u \subset v$, since E is a $B(\mathscr{A})$ group. Thus, $u \supset v$ and $u \subset v$ imply $u = v$.

For $B_r(\mathscr{A})$ groups it is easy to check that the condition in Theorem 4 is also sufficient.

32.   THE OPEN HOMOMORPHISM AND CLOSED GRAPH THEOREMS

As was pointed out in the beginning of this chapter, whenever the statements (o.h.) and (c.g.) are true, they are called the open homomorphism and closed graph theorems, respectively.

The two statements (o.h.) and (c.g.) are not generally true, as the following example demonstrates.

Example.   Let $E = R$, the real line endowed with the discrete topology and $F = R$, the real line endowed with the usual Euclidean metric topology. As is shown in §19, Chapter III, Examples 1 and 8, E and F are Hausdorff topological additive abelian groups. Now consider the identity homomorphism $i : E \to F$. Then i is a continuous $1:1$ homomorphism of E onto F. Since the discrete topology is strictly finer than the Euclidean topology R, i cannot be open.

Furthermore let $j : F \to E$, where $j = i^{-1}$, the inverse of i. Clearly j is an isomorphism of F onto E. Since i is continuous, the graph of i is closed in $E \times F$. The graph of j, being the same as that of i, is also closed in $F \times E$. But j cannot be continuous, because otherwise i would be open which is not true.

Thus, this simple example shows the following:

Proposition 10. (a) *A continuous homomorphism of a locally compact Hausdorff topological group into another locally compact Hausdorff topological group (or a metrizable complete topological group) is not necessarily open.*

(b) *A homomorphism f of a locally compact Hausdorff (or metrizable complete) group onto another locally compact Hausdorff group such that the graph of f is closed in the product space, is not necessarily continuous.*

The above example thus reveals that for the statements (o.h.) and (c.g.) to be true for locally compact groups E and F extra hypotheses are needed. In the sequel, we prove a very general theorem from which almost all existing open homomorphism and closed graph theorems for topological groups can be derived.

Theorem 5. (*Closed graph.*) *Let $\mathscr{C}$ be a class of Hausdorff topological groups satisfying the following: For an arbitrary Hausdorff topological group H if there exists a continuous almost open homomorphism of some member in $\mathscr{C}$ into H, then H also belongs to $\mathscr{C}$. Let $F \in \mathscr{C}$ and let E_u be a $B_r(\mathscr{C})$ group. Let f be an almost continuous and almost open homomorphism of F into E such that the graph of f is closed in $F \times E_u$. Then f is continuous.*

PROOF. Let $\{U\}$ denote a fundamental system of closed symmetric neighborhoods of e in E and let $\{U\}$ satisfy the conditions of Theorem 3, §20, Chapter III. For each U in $\{U\}$, define

$$\overset{*}{U} = \overline{f(\overline{f^{-1}(U)})}.$$

Clearly, each $\overset{*}{U}$ contains e and is closed symmetric (Lemma 2, §30). Now to show that the condition (b) of Theorem 3, §20, Chapter III, is satisfied by $\{\overset{*}{U}\}$, let U_1 be in $\{U\}$ such that $U_1 U_1 \subset U$. Then

$$\overset{*}{U} \supset \overline{f(\overline{f^{-1}(U_1 U_1)})} \supset \overline{f(\overline{f^{-1}(U_1)f^{-1}(U_1)})}$$
$$\supset \overline{f(\overline{f^{-1}(U_1)})f(\overline{f^{-1}(U_1)})} = \overset{*}{U_1}\overset{*}{U_1},$$

by Lemmas 1 and 2, §30. Furthermore, for each U in $\{U\}$ there exists U_1 such that $a^{-1}Ua \supset U_1$ or $U \supset aU_1a^{-1}$ for any fixed $a \in G$. Thus, again using Lemmas 1 and 2, §30, one has the following:

$$\overset{*}{U} = \overline{f(\overline{f^{-1}(U)})} \supset \overline{f(\overline{f^{-1}(aU_1a^{-1})})}$$
$$\supset \overline{f(\overline{f^{-1}(a)f^{-1}(U_1)f^{-1}(a^{-1})})}$$
$$\supset \overline{f(\overline{f^{-1}(a)})f(\overline{f^{-1}(U_1)})f(\overline{f^{-1}(a^{-1})})}$$
$$\supset a\overset{*}{U_1}a^{-1}$$

or $a^{-1}\overset{*}{U}a \supset \overset{*}{U_1}$. This proves (c) of Theorem 3, §20, Chapter III. Therefore, there exists a topology v on E having $\{\overset{*}{U}\}$ as a basis of neighborhoods of e and such that E_v is a topological group.

Now to show that E_v is a Hausdorff space, we use the condition that the graph of f is closed. For each U in $\{U\}$, there exists U_1 in $\{U\}$ such that $U_1 U_1 \subset U$. This shows that $\overset{*}{U_1}\overset{*}{U_1} \subset \overset{*}{U}$, as proved above. Let $y \in \overset{*}{U}$ for each $\overset{*}{U}$ in $\{\overset{*}{U}\}$. Then $y \in \overset{*}{U_1}$. Hence, there exists $x_1 \in \overline{f^{-1}(U_1)}$ such that $f(x_1) \in yU_1$. But $\overline{f^{-1}(U_1)} \subset Vf^{-1}(U_1)$, where V is an arbitrary neighborhood of e' in F, owing to Proposition 3, §20, Chapter III. Therefore, $x_1 \in Vf^{-1}(U_1)$. This means that there exists $x_2 \in V$ such that $x_2^{-1}x_1 \in f^{-1}(U_1)$. Since f is a homomorphism, it follows that

$$[f(x_2)]^{-1}f(x_1) \in U_1,$$

or

$$f(x_2) \in f(x_1)U_1 \text{ (since } U_1 \text{ is symmetric).}$$

Hence,

$$f(x_2) \in yU_1 U_1 \subset yU.$$

In other words,

$$(x_2, f(x_2)) \in (V, yU) \cap G,$$

where G is the graph of f. Since (V, yU) is an arbitrary neighborhood of (e', y), $(e', y) \in \bar{G} = G$, because G is closed by hypothesis. Hence, $e = f(e') = y$. In other words, $\cap \overset{*}{U} = \{e\}$. Thus, by Theorem 4, §21, Chapter III, E_v is a Hausdorff topological group and therefore regular by Theorem 5, §21, Chapter III.

For each $\overset{*}{U}$, clearly $\overset{*}{U} \supset U$ and, therefore, $u \supset v$. Hence by Proposition 8, §31, $u \supset v(u) \supset v$. Now we wish to show that $v(u) = v$.

Let $y \in \overset{*}{U}$. Then we show that $y \in Cl_vU$. Let W, W_1 be in $\{U\}$ such that $W_1 W_1 \subset W$. Then $\overset{*}{W_1}\overset{*}{W_1} \subset \overset{*}{W}$, as shown above. Now $y \in \overset{*}{U}$ implies $yW_1 \cap f(\overline{f^{-1}(U)}) \neq \varnothing$; that is, there exists $x_1 \in \overline{f^{-1}(U)}$ such that $f(x_1) \in yW_1$. Since f is almost continuous by hypothesis, $\overline{f^{-1}(W_1)}$ is a neighborhood of e' in F. Hence, by Proposition 3, §20, Chapter III, $x_1 \in \overline{f^{-1}(U)}$ implies

$$x_1 \in f^{-1}(U)\overline{f^{-1}(W_1)}.$$

That means there exists $x_2 \in f^{-1}(U)$ such that

$$x_2^{-1}x_1 \in \overline{f^{-1}(W_1)}.$$

Since f is a homomorphism,

$$[f(x_2)]^{-1}f(x_1) \in f(\overline{f^{-1}(W_1)}) \subset \overset{*}{W_1}.$$

But $\overset{*}{W_1}$ being symmetric, we have

$$[f(x_1)]^{-1}f(x_2) \in \overset{*}{W_1},$$

or

$$f(x_2) \in f(x_1)\overset{*}{W_1} \subset y\overset{*}{W_1}\overset{*}{W_1} \subset y\overset{*}{W_1}\overset{*}{W_1} \subset y\overset{*}{W}.$$

Since $x_2 \in f^{-1}(U)$, $y\overset{*}{W} \cap U \neq \varnothing$. But $\overset{*}{W}$ being arbitrary, $y \in Cl_v U$. This shows that $v(u) \subset v$. Now combining the two inclusion relations between v and $v(u)$, we have $v = v(u)$. Thus, so far we have shown that $u \supset v$ and $v = v(u)$ on E.

Since for each $\overset{*}{U}$ in $\{\overset{*}{U}\}$,

$$f^{-1}(\overset{*}{U}) = f^{-1}(\overline{f(\overline{f^{-1}(U)})}) \supset \overline{f^{-1}(U)}$$

and since, f being almost continuous, $\overline{f^{-1}(U)}$ is a neighborhood of e' in F, it follows that the mapping $f: F \to E_v$ is continuous. Further, for each neighborhood V of e' in F,

$$Cl_v f(V) \supset Cl_u f(V)$$

since $v \subset u$. Therefore, $f: F \to E_v$ is almost open because $f: F \to E_u$ is almost open by hypothesis. Hence, $f: F \to E_v$ being a continuous and almost open mapping of $F \in \mathscr{C}$ into E_v, it follows by the assumption on $\mathscr{C}$ that $E_v \in \mathscr{C}$. But since E_u is a $B_r(\mathscr{C})$ group and the identity mapping $i: E_u \to E_v$ is continuous and almost open (because $u \supset v$ and $v(u) = v$), it follows that i is open, i.e., $u = v$. Since $f: F \to E_v$ has been proved to be continuous, so is $f: F \to E_u$. This proves the theorem.

For $B_r(\mathscr{A})$ groups the hypothesis of the above theorem can be weakened, as is the case in the following:

Theorem 6. *Let F be any Hausdorff topological group and E_u a $B_r(\mathscr{A})$ group. Let f be an almost continuous homomorphism of F into E, the graph of which is closed in $F \times E$. Then f is continuous.*

PROOF. The topology v on E defined in the above theorem is a Hausdorff topology (since the graph of f is closed in $F \times E$), and $u \supset v$. Also E_v is a topological group. Moreover, since f is almost continuous, the identity mapping $i: E_u \to E_v$ is almost open because $v(u) = v$, as shown above. Thus, since E_u is a $B_r(\mathscr{A})$ group, it follows that i is open or $u = v$. Since $f: F \to E_v$ is continuous, as shown above, so is $f: F \to E_u$.

The topology v discussed in Theorems 5 and 6 was first introduced by A. P. Robertson and W. Robertson[40] in connection with a closed graph theorem for a class of locally convex spaces. However, Theorem 5 for locally convex spaces is due to the author[21] and Theorem 6 is due to Pták.[38]

Theorem 7. *Let F be a compact Hausdorff topological group and E a $B_r(\mathcal{K})$ group, where $\mathcal{K}$ is the class of all compact Hausdorff topological groups. Let f be an almost continuous homomorphism of F onto E_u. If the graph of f is closed in $F \times E$, then f is continuous.*

PROOF. As in Theorem 5, the topology v on E is Hausdorff and $u \supset v$. Since F is compact and $f: F \to E_v$ is continuous, E_v is also compact. Since $u \supset v$ and $v(u) = v$, as proved in Theorem 5, the identity mapping: $E_u \to E_v$ is continuous and almost open. Now since E_u is a $B_r(\mathcal{K})$ group, it follows that $u = v$. This proves that f is continuous.

Theorem 8. (*Open homomorphism.*) *Let E be a $B_r(\mathcal{A})$ group and F any Hausdorff topological group. Let g be a homomorphism of E onto F such that the graph of g is closed. Then g is open if g is almost open.*

PROOF. Let $H = \{x \in E : g(x) = e' \in F\}$. Then H is obviously an invariant subgroup of E. By Proposition 2, §30, H is a closed subgroup since the graph of g is closed by hypothesis. Now H being an invariant closed subgroup of E, E/H is a $B_r(\mathcal{A})$ group by Proposition 7, §31. Let $\varphi: E \to E/H$ be the canonical mapping and $f: E/H \to F$. Then $g = f \circ \varphi$. To see that the graph of f is closed in $(E/H) \times F$, let G_f and G_g denote the graphs of f and g in $(E/H) \times F$ and $E \times F$, respectively. Let $(\dot{x}, y) \in \bar{G}_f$, where $\dot{x} \in E/H$, $y \in F$. For each neighborhood $\dot{U}$ of $\dot{x} = xH$ and V of y, $(\dot{U}, V) \cap G_f \neq \varnothing$. Now for any neighborhood W of x, $\dot{W}$ being a neighborhood of $\dot{x}$, $(\dot{W}, V) \cap G_f \neq \varnothing$. Therefore, there exists a $t \in W$ such that $\dot{t} \in \dot{W}$ and $f(\dot{t}) \in V$. But $f(\dot{t}) = g(t)$ and so $(t, g(t)) \in (W, V) \cap G_g \neq \varnothing$. This shows that $(x, y) \in \bar{G}_g = G_g$, because G_g is closed by hypothesis. Hence, $y = g(x) = f(\varphi(x)) = f(\dot{x})$. In other words, $(\dot{x}, y) = (\dot{x}, f(\dot{x})) \in G_f$. Therefore, G_f is closed.

Furthermore, f is almost open because g is (Proposition 3, §30). Moreover, f is $1:1$ and onto. Therefore, f^{-1} exists and its graph is closed because it is equal to G_f, which is closed. Also f^{-1} is almost continuous because f is almost open. Hence, by Theorem 6, f^{-1} is continuous. This means that f is open. Since φ is open, this proves that g is open.

Theorem 9. *Let E be a $B_r(\mathcal{K})$ group and F a compact Hausdorff topological group, where $\mathcal{K}$ is the class of all compact Hausdorff groups. Let f be an almost open homomorphism of E onto F. If the graph of f is closed in $E \times F$, then f is open.*

PROOF. Since the graph of f is closed, $f^{-1}(e')$ is a closed invariant subgroup of E. By Proposition 7, §31, $E/f^{-1}(e')$ is also a $B_r(\mathcal{K})$ group because E is so. Now, as in Theorem 8, it can be assumed that f is $1:1$. Thus, the inverse f^{-1} of f exists and is almost continuous because f is almost open. Furthermore, the graph of f^{-1}, being the same as that of f, is closed. Hence, by Theorem 7, f^{-1} is continuous or f is open.

Proposition 11. *Let E be a separable (or Lindelöf) Hausdorff topological group and F a Baire group. Then*

(a) *Each homomorphism f of E onto F is almost open.*

(b) *Each homomorphism g of F into E is almost continuous.*

PROOF. (a) Assume first that E is separable. Let $\{x_n\}$ be a countable dense subset of E. Let U and U_1 be symmetric neighborhoods of e in E so that $U_1^2 \subset U$. Then clearly $\bigcup_{n\geq 1} x_n U_1 = E$, and $F = f(E) = \bigcup_{n\geq 1} f(x_n)f(U_1)$. Since F is a Baire group, at least for one n, $\overline{f(x_n)f(U_1)} = f(x_n)\overline{f(U_1)}$ has an interior point. Since $\overline{f(U_1)}$ and $f(x_n)\overline{f(U_1)}$ are homeomorphic, $\overline{f(U_1)}$ has an interior point y_0, say. By the symmetry of U_1, in view of Lemma 2, §30, $\overline{f(U_1)} = [\overline{f(U_1)}]^{-1}$ and, thus, $e' = y_0 y_0^{-1}$ is an interior point of $\overline{f(U_1)}\,\overline{f(U_1)} \subset \overline{f(U_1)f(U_1)} = \overline{f(U_1^2)} \subset \overline{f(U)}$. Hence, f is almost open.

Assume E is Lindelöf. For each neighborhood U of e, the family $\{xU\}$ covers E. Then a countable subfamily $\{x_n U\}$ of $\{xU\}$ will cover E. Now going through the same steps as above, one proves that f is almost open.

(b) Suppose E is Lindelöf. Let U and V be neighborhoods of e in E such that $U^2 \subset V$. Then $g(F) \subset \overline{g(F)} \subset Ug(F)$ and so only a countable subfamily $\{Ug(x_n)\}$, say, covers $g(F)$ because every closed subspace of a Lindelöf space is a Lindelöf space. Since $\bigcup_{n\geq 1} g^{-1}(Ug(x_n)) = F$, the sets of $\{Ug(x_n)\}$ are all homeomorphic and F is a Baire space, it follows that $\overline{g^{-1}(U)}$ has an interior point, say x_0. But then $e = x_0 x_0^{-1}$ is an interior point of

$$\overline{g^{-1}(U)}\,\overline{g^{-1}(U)} \subset \overline{g^{-1}(U)g^{-1}(U)} \subset \overline{g^{-1}(V)}$$

(Lemma 2, §30). This shows that g is almost continuous. Similarly, one proves the proposition when E is separable.

See Kelley,[27] p. 213, for similar ideas.

Proposition 12. *Proposition 11 is true for the following pairs E and F of Hausdorff topological groups.*

(a) *E compact and F a Baire space*

(b) *E compact and F locally compact or metrizable complete*

(c) *E and F both compact.*

PROOF. Since a compact space is a Lindelöf as well as a Baire space, and a locally compact, metrizable, complete space is a Baire space, Proposition 12 follows from Proposition 11.

Corollary 3. *Let E be a separable (or Lindelöf) B($\mathscr{A}$) group and F a Baire Hausdorff topological group. Then*

(o.h.) *Each continuous homomorphism f of E onto F is open.*

(c.g.) *Each homomorphism g of F into E, the graph of which is closed in F $\times$ E, is continuous.*

PROOF.   By Proposition 11, f is almost open and g is almost continuous. Hence (o.h.) follows by the definition of $B(\mathscr{A})$ groups, and (c.g.) follows by Theorem 6.

Corollary 4.   *Let E be a separable metric complete semitopological group or a Lindelöf metric complete topological group, and F a Baire Hausdorff topological group. Then the statements (o.h.) and (c.g.) are true.*

PROOF.   This is a particular case of Corollary 3 since E, being a separable metric complete semitopological group, is a $B(\mathscr{A})$ group by Corollary 2, §31.

A particular case of Corollary 4 is:

Corollary 5.   *Let E be a complete Hausdorff topological group satisfying the second axiom of countability, and F a Baire Hausdorff topological group. Then the statements (o.h.) and (c.g.) are true.*

Corollary 6.   *Let E be a separable complete metric semi-topological group or a Lindelöf complete metric topological group, and F a metrizable complete topological group. Then (o.h.) and (c.g.) are true.*

PROOF.   This follows from Corollary 4, because F is a Baire space by a well-known result of Baire (Theorem 1, §2, Chapter I).

Corollary 7.   *Let E be a locally compact Hausdorff topological group satisfying the second axiom of countability, and F a Baire (or metrizable complete) group. Then (o.h.) and (c.g.) are true.*

PROOF.   By Corollary 2 (Chapter IV, §26), E is complete metrizable and therefore separable because E satisfies the second axiom of countability. Thus the corollary follows from Corollary 5 or 6.

Corollary 8.   *Let E be a locally compact regular semitopological group satisfying the second axiom of countability, and F a locally compact Hausdorff topological group. Then (o.h.) and (c.g.) are true.*

PROOF.   This is a particular case of Corollary 7 since F, being locally compact, is a Baire space and E is a topological group by Corollary 5, §17, Chapter II.

Corollary 9.   *Let E and F be both regular locally compact semitopological groups satisfying the second axiom of countability. Then the statements (o.h.) and (c.g.) are true.*

PROOF.   By Corollary 5, §17, Chapter II, E and F are topological groups, and hence this is a particular case of Corollary 8.

Corollary 10. *Let E be a Lindelöf (or separable) B($\mathscr{A}$) group and F a locally compact Hausdorff topological group. Then (o.h.) and (c.g.) are true.*

PROOF. Since a locally compact Hausdorff topological group is a Baire space, this corollary is a particular case of Corollary 3.

The following corollary is a particular case of the above:

Corollary 11. *Let E be a Lindelöf (or separable) metrizable complete semitopological group and F a locally compact Hausdorff topological group. Then (o.h.) and (c.g.) are true.*

Corollary 12. *Let E be a compact Hausdorff semitopological group and F any topological group. Then (o.h.) is true. Furthermore (c.g.) is true provided F is a Baire group.*

PROOF. By Exercise B (1), Chapter II, E is a topological group. Thus the proof of the statement (o.h.) is contained in Theorem 2 (§31). For (c.g.), we observe that every compact space is Lindelöf. Therefore $g: F \to E$ is almost continuous by Proposition 12. Now in view of Theorem 2, the corollary follows from Theorem 6.

The following corollaries are particular cases of the above:

Corollary 13. *Let E be a compact Hausdorff semitopological group and F a locally compact Hausdorff topological group. Then (o.h.) and (c.g.) are true.*

Corollary 14. *Let E and F both be compact Hausdorff semitopological groups. Then (o.h.) and (c.g.) are true.*

Exercises

1. Let E_u be a Hausdorff topological group. E_u is said to be a *minimal* group if there exists no Hausdorff topology compatible with the group structure and strictly coarser than u.

(a) Show that each compact Hausdorff topological group is minimal.

(b) Show that every minimal group is a $B_r(\mathscr{A})$ group.

(For a more general concept of minimality for linear spaces see Husain.[25])

2. Let F be any Hausdorff topological group and E a minimal group. Show that

(a) A $1:1$ continuous homomorphism of E onto F is open.

(b) A homomorphism of F into E, the graph of which is closed in $F \times E$, is continuous.

3. Is every $B(\mathscr{A})$ group complete? (This is an open question.)

4. Show that every locally compact Hausdorff topological group is a $B(\mathscr{A})$ group.

VI

Haar Measure

In this chapter we shall discuss a generalization—Haar measure—of Lebesgue measure on the real line. Haar measure was first introduced by A. Haar[15] in 1933. On a locally compact Hausdorff topological group, Haar measure is the analogue of Lebesgue measure on the real line. Indeed, on the real line Haar measure coincides with Lebesgue measure. We note that there are other generalizations of Lebesgue measure on the real line although they will not concern us here. The material contained herein is standard. A recent exhaustive treatment of the subject can be found in the book of Hewitt and Ross,[19] and in other valuable books e.g., Loomis,[29] Pontrjagin,[37] and Weil.[46] The proof given here of the existence and "essential" uniqueness of Haar measure on a locally compact Hausdorff topological group, is due to Weil.[46] The proof of Cartan, which avoids the Tychonoff theorem, is also given.

Throughout this chapter we shall assume that a locally compact topological group is Hausdorff. Moreover, it is also assumed that the reader is familiar with the theory of Lebesgue measure, available, e.g., in the book of Halmos[16] and in others.

As is known, every regular measure μ on a measure space gives rise to an integration theory. Moreover, the integration process is linear, i.e., if two functions can be integrated then their sum as well as their scalar multiples can be integrated. Conversely, given an integration process on a suitable class of real or complex functions defined on a set X, one can define a measure on a certain class of subsets of X. We shall actually see briefly in the sequel how an integration process leads to a measure. Incidentally, it should be remarked that although integration is linear, a linear functional is not necessarily an integral. Indeed, the relationships among linear functionals, integrals, and measures go much deeper than the discussion given here.

In §§33, 34, we survey some basic facts about integration, and in §§35, 36, the existence and "uniqueness" of Haar integral on locally compact groups are established.

33. MEASURE AND INTEGRATION ON LOCALLY COMPACT SPACES

As remarked above, we assume that the reader is familiar with Lebesgue theory on the real line. However, to facilitate the understanding of the forthcoming sections of this chapter, we should have a survey of measure and integration theory on a locally compact Hausdorff space. The treatment of the material is by no means exhaustive. Only the relevant points are taken up. For a detailed discussion, the reader is referred to the books mentioned above.

The main purpose of this section is to introduce quickly the inter-connection of linear functionals and integrals on a certain class of functions defined on a locally compact Hausdorff space.

Let X denote a locally compact Hausdorff space. Let $B(X)$ and $C(X)$ denote, respectively, the set of all bounded and continuous bounded real- (or complex-) valued functions. Let $C_0(X)$ denote the set of all continuous real- (or complex-) valued functions on X, each of which vanishes outside a compact set of $\bar{X}$. This compact set is not necessarily the same for all functions. It is clear that $B(X)$, $C(X)$, and $C_0(X)$ form linear spaces, and $C_0(X) \subset C(X) \subset B(X)$. Let $C_0^+ = \{f \in C_0(X) : f \geq 0\}$. We shall discuss only real functions in the sequel. For a complex function $f = f_1 + if_2$, $i = \sqrt{-1}$, the arguments used in the sequel can be applied for f_1 and f_2 to produce desired conclusions for f.

Definition 1. (a) Let E be a real linear space. A real-valued function f on E is said to be a *functional*.

(b) A functional is said to be *additive* (or *subadditive*) if for each pair $x, y \in E$, $f(x + y) = f(x) + f(y)$ (or $f(x + y) \leq f(x) + f(y)$).

(c) A functional is said to be *positive homogeneous* if for each $x \in E$ and real $\lambda \geq 0$, $f(\lambda x) = \lambda f(x)$. If for each real λ and $x \in E$, $f(\lambda x) = \lambda f(x)$, then f is said to be *homogeneous*.

(d) A homogeneous additive functional is said to be *linear*. A linear functional I on a linear space E is *non-negative* if $I(f) \geq 0$ for all $f \geq 0$

(e) Let L be any linear subspace of $B(X)$ such that L is closed under *lattice operations*, viz., for $f, g \in L$, $\max(f, g) = f \vee g$ and $\min(f, g) = f \wedge g$ (max and min are taken pointwise) are also in L.

A linear functional I on L is said to be an *integral* (or more specifically a *Daniell integral*) if I satisfies the following properties:

(i) For $f, g \in L$, $f \geq g$ implies $I(f) \geq I(g)$.

(ii) For each monotonically decreasing sequence $\{f_n\}(n \geq 1)$, $f_n \in L$ that converges to the zero function pointwise, $I(f_n)$ converges to 0, where $I(f_n)$ is a monotonically decreasing sequence of real numbers. From (i) it follows that an integral is non-negative.

The Riemann integral $\mathscr{R} \int f(x)\,dx$, $f(x) \in C$ $[0, 1]$ is an integral in the above sense, since pointwise monotone (decreasing) convergence implies uniform convergence on $[0, 1]$ (cf., the proof of Theorem 1 below).

By definition, an integral is a linear functional. The following theorem gives the converse under suitable conditions.

Theorem 1. *Let X be a locally compact Hausdorff space and $C_0(X)$ as above. Then every non-negative linear functional I on $C_0(X)$ is an integral.*

PROOF. Condition (i) in the definition of an integral is clearly satisfied. To show (ii), we first prove the following:

(a) Let $\{f_n\} \subset C_0(X)$ be a pointwise monotonically decreasing sequence, i.e., $f_{n+1}(x) \le f_n(x)$ for each $n \ge 1$ and for all $x \in X$. Suppose $f_n \to 0$ pointwise. For an arbitrary $\varepsilon > 0$, let $E_n = \{x \in X : f_n(x) \ge \varepsilon\}$. Since f_n is continuous and vanishes outside a compact set of X, E_n, being a closed subset of the support of f_n, is compact for each $n \ge 1$. Moreover, since $f_n(x) \to 0$ for each x, $\bigcap_{n \ge 1} E_n = \varnothing$. Now compactness of E_n implies there is at least one m for which $E_m = \varnothing$. Hence $\sup_x |f_n(x)| \le \varepsilon$ for all $n \ge m$. This proves that $f_n \to 0$ in the sup norm topology on $C_0(X)$.

(b) Now we show that I is uniformly bounded on a subspace $C_0(A)$ of $C_0(X)$, where A is a compact subset of X and $C_0(A)$ consists of those $f \in C_0(X)$ that vanish outside A. For this, let $g \in C_0^+(X)$ such that $g(x) \ge 1$ for all $x \in A$. Then for any $f \in C_0(A)$, $|f(x)| \le \sup_{x \in X} |f(x)| g(x)$. Hence, since I is positive-homogeneous,

$$I(f) \le \sup_{x \in X} |f(x)| I(g),$$

and

$$\|I\| = \sup_{\|f\| \le 1} |I(f)| \le \sup_{x \in X} |f(x)| I(g) < \infty.$$

This shows that I is bounded on $C_0(A)$.

Finally, let $f_n \to 0$, where $\{f_n\}$ is a monotonically decreasing sequence and the convergence is pointwise. Then, as proved above, pointwise convergence implies $\sup_x |f_n(x)| \to 0$ when $n \to \infty$, where $\sup_x |f_n(x)|$ is a monotonically decreasing sequence of real numbers. If $f_1 \in C_0(A)$, then $f_n \in C_0(A)$ for all $n \ge 1$. Let $M > 0$ such that $\|I\| \le M$ on $C_0(A)$ by (b). Then

$$I(f_n) \le M \sup_x |f_n(x)| \to 0 \qquad \text{as} \quad n \to \infty$$

proves that I is an integral.

Having defined an integral on a linear subspace L of $B(X)$, one tries to extend this integral to a larger class of functions that includes L. This extension evolves in the following steps:

Step 1. Let $M(L)$ denote the set of all functions that are limits of monotonic sequences of functions in L. If ∞ is permitted to be a possible value, then every increasing sequence of functions is convergent. Clearly L is a subset of $M(L)$, since $\{f_n = f\}$ for all $n \geq 1, f \in L$, is such a sequence. Thus, the obvious extension of I from L to $M(L)$ is

$$I(f) = \lim_{n \to \infty} I(f_n),$$

where $f_n \in L$ and $\{f_n\}$ is an increasing sequence that converges to f.

Step 2. One shows that the extension of I is independent of choice of the sequence, i.e., if $\{f_n\}$ and $\{g_n\}$ are two increasing sequences such that $f_n \to f$ and $g_n \to f$, then

$$I(f) = \lim_{n \to \infty} I(f_n) = \lim_{n \to \infty} I(g_n).$$

Step 3. Let $M_-(L) = \{f: -f \in M(L)\}$. Then one defines I on $M_-(L)$ as follows:

$$I(f) = -I(-f).$$

If $f \in M(L) \cap M_-(L)$, then the two definitions agree.

Step 4. Finally, one defines the class L_1, the set of all I-summable functions. f is said to be *I-summable* if for each $\varepsilon > 0$ there exist $g \in M_-(L)$ and $h \in M(L)$ such that $g \leq f \leq h$, $I(g)$ and $I(h)$ are finite, and $|I(h) - I(g)| < \varepsilon$.

The extension of I, also denoted by I, to L_1 is defined by the following: for $f \in L_1$, $I(f) = \inf_{h \in M(L)} I(h) = \sup_{g \in M_-(L)} I(g)$.

It is easy to verify that L_1 is a vector space and I is a positive linear functional on L_1. The fact that I is an integral on L_1 follows from the more general theorem below.

Theorem 2. (*Monotone Convergence Theorem.*) *If $\{f_n\} \subset L_1$ is a mono-tonically increasing sequence converging to f pointwise such that $\lim_{n \to \infty} I(f_n) < \infty$, then $f \in L_1$ and $\lim_{n \to \infty} I(f_n) = I(f)$.*

PROOF. First we prove the theorem when $\{f_n\} \subset M(L)$. Let $\{g_{nm}\}(m > 1)$ be a sequence in L converging to f_n for each $n \geq 1$. (Such sequences exist by definition of $M(L)$.) Since L is closed under the lattice operations, $h_n = g_{1n} \vee g_{2n} \vee \cdots \vee g_{nn} \in L$; and $\{h_n\}$ is an increasing sequence. Clearly, $g_{mn} \leq h_n \leq f_n$ for $m < n$. Hence, $f = \lim_{m \to \infty} f_m = \lim_{m \to \infty} (\lim_{n \to \infty} g_{mn}) \leq \lim_{n \to \infty} h_n \leq \lim_{n \to \infty} f_n = f$. This shows that $\lim_{n \to \infty} h_n = f \in M(L)$. Also the inequality $g_{mn} \leq h_n \leq f_n$ for $m < n$, implies $I(g_{mn}) \leq I(h_n) \leq I(f_n)$ for $m < n$. Whence, $\lim_{m \to \infty} I(f_m) \leq I(f) \leq \lim_{n \to \infty} I(f_n)$ proves the theorem when $\{f_n\} \subset M(L)$.

Now let $\{f_n\} \subset L_1$. Let $\varepsilon > 0$ be given. Without loss of generality we can assume that $f_1 = 0$. By the definition of L_1, there exists $h_n \in M(L)$ for each $n \geq 1$ such that $f_{n+1} - f_n \leq h_n$ and $I(h_n) < I(f_{n+1} - f_n) + \varepsilon/2^n$. Putting $h = \sum_{n=1}^{\infty} h_n$, we have $h \in M(L)$ and $I(h) = \sum_{n=1}^{\infty} I(h_n)$ by the previous paragraph. Clearly, $f \leq h$ and $I(h) \leq \lim_{n \to \infty} I(f_n) + \varepsilon$. Similarly, if n is large enough we can find $g \in M_-(L)$ such that $g \leq f_n \leq f \leq h$ and $I(h) - I(g) < 2\varepsilon$. This shows that $f \in L_1$ and $\lim_{n \to \infty} I(f_n) = I(f)$. This completes the proof.

Very often, to avoid mentioning the validity of a property outside a set of measure zero, one defines the class L_1 of summable functions as the intersection of what we have called L_1 and $\mathscr{B}$, where $\mathscr{B}$ is the smallest family containing L and such that the limit of each increasing as well as each decreasing sequence in L is in $\mathscr{B}$. Each member f of $\mathscr{B}$ is said to be a *Baire function.*

One can show that if $f \in \mathscr{B}$ then there exists $g \in M_-(L)$ such that $f \leq g$. Let $\mathscr{B}^+ = \{f \in \mathscr{B} : f \geq 0\}$. Then it follows that $M(L^+) = \mathscr{B}^+$, where $M(L^+)$ is the smallest family containing L^+ and the limit of each increasing as well as decreasing sequence in L^+. Now in order that $f \in L_1 \cap \mathscr{B}$, it is necessary and sufficient that there exists $g \in L_1$ such that $|f| \leq g$. From this it is possible to extend I to the whole of $\mathscr{B}^+$. (cf. Loomis,[29] pp. 33–34.)

A function $f \in \mathscr{B}$ is said to be *I-integrable* or just simply *integrable* if both $f^+ \in L_1$ and $f^- \in L_1$, where

$$f^+ = \begin{cases} f & \text{if } f \geq 0 \\ 0 & \text{if } f < 0 \end{cases}$$

$$f^- = \begin{cases} -f & \text{if } f \leq 0 \\ 0 & \text{if } f > 0. \end{cases}$$

It is clear that $f = f^+ - f^-$ and $|f| = f^+ + f^-$.

Step 5. Let A be a subset of X and let χ_A denote the *characteristic function* of A, i.e., $\chi_A(x) = 1$ or 0 according as $x \in A$ or $x \in A$. For $\chi_A \in \mathscr{B}$, define

(1) $$\mu(A) = I(\chi_A).$$

Then the set function μ satisfies the following properties:

(i) For integrable χ_A, $0 \leq \mu(A) \leq \infty$

(ii) $\mu(\varnothing) = I(\chi_\varnothing) = 0$

(iii) For a sequence $\{\chi_{A_n}\}$ of integrable functions such that $\{A_n\}$ is a sequence of pairwise disjoint sets,

$$\mu\left(\bigcup_{n=1}^{\infty} A_n\right) = \sum_{n=1}^{\infty} \mu(A_n) = \sum_{n=1}^{\infty} I(\chi_{A_n}).$$

(a) Let **R** denote a nonempty family of subsets of a set X. **R** is said to be a *ring* (or *Boolean ring*) if for A, $B \in$ **R**, $A \cup B$, and $A \sim B$ are in **R**.

It is easy to check that a ring is also closed under finite intersections.

(b) A ring **R** is said to be a *σ-ring* if for each sequence $\{A_n\}$ in **R**, $\bigcup_{n=1}^{\infty} A_n \in$ **R**.

(c) A real-valued function m defined on a ring **R** is said to be a *measure* if

(i) $m(\varnothing) = 0$, $\varnothing$ empty set

(ii) $0 \leq m(A) \leq \infty$ for $A \in$ **R**

(iii) For each pairwise disjoint sequence $\{A_n\}$ in $\mathbf{R}$, $\bigcup\limits_{n=1}^{\infty} A_n \in \mathbf{R}$ implies

$$m\left(\bigcup_{n=1}^{\infty} A_n\right) = \sum_{n=1}^{\infty} m(A_n).$$

This is the usual but restricted definition of measure following Halmos.[16] For a more general definition, one omits (ii) in (c).

From Step 5, it follows that μ defined by (1) is a measure on the class $\mathbf{R}(\mu)$ of sets A for which $\chi_A \in \mathscr{B}$. It can be shown that $\mathbf{R}(\mu)$ is a ring.

Let X be a locally compact Hausdorff space. Let $\mathbf{C}$ denote the family of all compact subsets of X. Let $\mathbf{S}$ denote the smallest σ-ring generated by $\mathbf{C}$, i.e., $\mathbf{S}$ is the intersection of all σ-rings containing $\mathbf{C}$. Then $\mathbf{S}$ is called the *Borel ring*, and each member of $\mathbf{S}$ is called a *Borel set*.

Let f be a real-valued function on X. f is said to be a *Borel-measurable* function if for each real number $a > 0$, $\{x \in X: |f(x)| < a\} \in \mathbf{S}$.

A complex- or real-valued set function defined on $\mathbf{S}$ is said to be a *Borel measure* if it satisfies (iii) above, and is finite on compact subsets in $\mathbf{S}$.

Let μ be a Borel measure on $\mathbf{S}$. Define the *total variation* $|\mu|$ of μ as follows:

$$|\mu|(A) = \sup\left\{\sum |\mu(A_n)| : \bigcup A_n \subset A\right\},$$

where sup is taken over all pairwise disjoint finite sequences $\{A_n\}$ in $\mathbf{S}$. Then it can be shown that $|\mu|$ is also a measure on $\mathbf{S}$. If, for each $E \in \mathbf{S}$,

$$|\mu|(E) = \sup |\mu|(K) = \inf |\mu|(V),$$

where K runs over compact subsets of E and V over all open Borel sets containing E, then μ us said to be a *regular measure*.*

Summing up, we have seen that an integral on a certain class of functions gives rise to a measure on a certain class of sets. Indeed, the reverse process is classical and well-known. Actually, it can be shown that the two processes are equivalent, (cf., Loomis[29]).

The rest of this section we devote to a very important theorem in the theory of integration, viz., the Radon-Nikodym theorem. From this we derive a useful result (Theorem 4) which will be used in Chapter VIII, §47.

A function is said to be an *I-null* function if $I(|f|) = 0$, where I is an integral a priori defined on L and then extended to $\mathscr{B}$. A set A is said to be *I-null* if its characteristic function χ_A is an *I*-null function. It is easy to see that $f \in \mathscr{B}$ is *I*-null if, and only if, the set $\{x: f(x) \neq 0\}$ is *I*-null.

Let I and J be two integrals. J is said to be *absolutely continuous* with respect to I, if every *I*-null set is *J*-null.

Before we prove the Radon-Nikodym theorem, we require the following:

Proposition 1. *Every bounded linear functional on L is the difference of two bounded, non-negative, positive-homogeneous, additive functionals.*

---

* The set of all regular measures forms a real vector space. If a measure is defined to be a complex-valued set function satisfying (iii) of (c), then the set of all regular measures forms a complex vector space. If $\|\mu\| = |\mu|(X) < \infty$, then μ is called a *bounded* measure and $\|\mu\|$ is the *norm* of μ.

PROOF. Let F be a bounded linear functional on L. For $0 \leq f \in L$, define $F^+(f) = \sup \{F(g): 0 \leq g < f\}$. Clearly, if $c > 0$, $F^+(cf) = cF^+(f)$, and $F^+(f) \geq 0$. To show that F^+ is additive, let $f_1, f_2 \geq 0$ (in L). If $0 \leq g_1 \leq f_1$ and $0 \leq g_2 \leq f_2$, then $0 \leq g_1 + g_2 \leq f_1 + f_2$ and $F^+(f_1 + f_2) \geq \sup F(g_1 + g_2) = \sup F(g_1) + \sup F(g_2) = F^+(f_1) + F^+(f_2)$. To prove the reverse inequality, let $0 \leq g \leq f_1 + f_2$ and choose $h \in L$ (for instance, $h = f_1 \wedge g$) such that $0 \leq h \leq f_1$, $0 \leq g - h \leq f_2$. Then $F^+(f_1 + f_2) = \sup F(g) \leq \sup F(h) + \sup F(g - h) \leq F^+(f_1) + F^+(f_2)$. The combination of two inequalities proves that F^+ is additive. Clearly, $|F^+(f)| \leq \|F\| \|f\|$ and hence F^+ is bounded, since F is by hypothesis. Furthermore, $F^+(f) \geq F(f)$. Thus, by putting $F^-(f) = F^+(f) - F(f)$, we see that F^+ and F^- satisfy the required conditions in the proposition.

Theorem 3. (*Radon-Nikodym.*) *Let I and J be two bounded integrals such that J is absolutely continuous with respect to I. Then there exists a unique I-summable function f_0 such that, for every J-summable f, ff_0 is I-summable and*

$$J(f) = I(ff_0).$$

PROOF. It is clear that $K = I + J$ is a bounded integral. Let $L_2(K)$ denote the set of functions f such that f^2 is K-summable. As in the classical case, $L_2(K)$ is a Hilbert space. For each $f \in L_2(K)$, $f = f \cdot 1 \in L_1(K)$ and by Schwarz's inequality (§11, Chapter I)

$$|J(f)| \leq J(|f|) \leq K(|f|) \leq \|f\|_2 \|1\|_2.$$

Hence, J is a bounded linear functional on $L_2(K)$, and therefore by Theorem 6, §11, Chapter I, there exists a unique $g \in L_2(K)$ such that

$$J(f) = \langle f, g \rangle = K(fg)$$

for each $f \in L_2(K)$, where $\langle f, g \rangle$ is the scalar product function for $L_2(K)$. Observe that $g \geq 0$ except on a K-null set. Repeating the argument, we have

$$J(f) = K(fg) = I(fg) + J(fg) = I(fg) + I(fg^2) + \cdots + I(fg^n) + J(fg^n)$$

$$= \sum_{m=1}^{n} I(fg^m) + J(fg^n) = I\left(f \sum_{n=1}^{n} g^m\right) + J(fg^n).$$

If we choose $f = \chi_A$, where $A = \{x : g(x) \geq 1\}$, then from this expansion it follows that A is I-null and hence J-null by hypothesis. This shows that fg^n converges to zero monotonically almost everywhere (i.e., everywhere except on a set of measure zero) for each $f \geq 0$. Since $f \in L_2(K) \subset L_1(J)$ (because J is a bounded linear functional on $L_2(K)$), by the monotone convergence theorem (Theorem 2), $J(fg^n)$ converges monotonically to zero.

Putting $f_0 = \sum_{m=1}^{\infty} g^m$, we see by the above expansion that ff_0 is I-summable and $J(f) = I(ff_0)$. In particular, taking $f = 1$, we see that f_0 is I-summable.

To show the uniqueness of f_0, we observe that $f_0 = g/(1 - g)$ or $g = f_0/(1 + f_0)$, which is unique for the representation of J. Since $J(f) = I(ff_0)$ for each $f \in L_2(K)$ which includes L, it follows that $J(f) = I(ff_0)$ for each J-summable f. For L is dense in $L_1(J)$ (by definition) and two bounded linear functionals coincide if they coincide on a dense subset. This completes the proof.

For $1 \le p < \infty$, define $L_p(I)$ as the set of all $f \in \mathscr{B}$ such that $|f^p|$ is I-summable. As in the classical case, it can be shown that $L_p(I)$ is a Banach space with the norm $\|f\|_p = \{I(|f|^p)\}^{1/p}$, $f \in L_p(I)$. It is of course understood that two functions are identified if they differ on a set of measure zero. In other words, $L_p(I)$ is the space of equivalence classes. However, for the sake of convenience we will suppress this precise connotation and will regard $L_p(I)$ as the space of functions. It can be shown that L is dense in $L_p(I)$, $1 \le p < \infty$.

By L_∞ we denote the space of all $f \in \mathscr{B}$ such that $|f|$ is bounded. (Indeed here again we identify functions differing on sets of measure zero.) L_∞ is also a Banach space with the norm: $\|f\|_\infty = \sup |f(x)|$.

It can be shown that if $f \in L_p$ $(p > 0)$ then $\lim_{q \to \infty} \|f\|_q$ exists, $(p^{-1} + q^{-1} = 1)$ and is equal to $\|f\|_\infty$.

Among L_p spaces, L_2 is most distinguished for the reason that L_2 is a Hilbert space.

Theorem 4. *Let I be a bounded integral and J a bounded linear nonzero functional on $L_p(I)$, $p \ge 1$. Then there exists a unique $f_0 \in L_q$ $(p^{-1} + q^{-1} = 1)$ if $p > 1$ $(f_0 \in L_\infty, q = \infty$ if $p = 1)$ such that $\|f_0\|_q = \|J\| = \sup_{\|f\|_p \le 1} |J(f)|$ and for each $f \in L_p(I)$*

$$J(f) = I(ff_0).$$

In other words, the dual (§11, Chapter I) of L_p $(p > 1)$ is L_q and the dual of L_1 is L_∞.

PROOF. By Proposition 1, $J = J^+ - J^-$ and each J^+, J^- is an integral on L, as is easy to check. If f is I-null then $\|f\|_p = 0$. Hence $J^+(f) = J^-(f) = 0$, which shows that J^+ and J^- are absolutely continuous with respect to I. By the Radon-Nikodym theorem there exists an I-summable function $f_0 \ne 0$ such that

$$J(f) = I(ff_0),$$

for each $f \in L_1(J^+) \cap L_1(J^-)$, and in particular for each $f \in L_p$, since $L_p \subset L_1(J^+) \cap L_1(J^-)$. If $p > 1$ and g any bounded function such that $0 \le g \le |f_0|$, then (putting $\varepsilon = |f_0|/f_0$)

$$I(g^q) \le I(g^{q-1}f_0\varepsilon) \le I(|g^{q-1}|) I(|f_0|) \le \|g^{q-1}\|_p \|J\|$$
$$\le \|J\| \{I(|g^{p(q-1)}|)\}^{1/p} = \|J\| \{I(|g^q|)\}^{1/p},$$

or $\|g\|_q = \{I(g^q)\}^{1/q} = \{I(g^q)\}^{1-(1/p)} \le \|J\|$. By choosing a monotonically increasing sequence $\{g_n\}$ such that $\{g_n\}$ converges to $|f_0|$ and $0 \leq g_n \leq |f_0|$, by the monotone convergence theorem and by the fact that L_q is complete, we conclude that $f_0 \in L_q$ and $\|f_0\|_q \leq \|J\|$.

Furthermore, by the Hölder inequality (§11, Chapter I, and Loomis,[29] p. 37) if $p > 1$,

$$|J(g)| \leq I(|gf_0|) \leq \|g\|_p \|f_0\|_q$$

and, hence, $\|J\| \leq \|f_0\|_q$. Combining the two inequalities, we obtain $\|J\| = \|f_0\|_q$, when $p > 1$. The uniqueness follows from the Radon-Nikodym theorem. The case when $p = 1$ is left for the reader to prove.

34. INTEGRATION ON PRODUCT SPACES AND FUBINI THEOREM

In this section, we see briefly how the integral on the functions of coordinate spaces gives rise to an integral for functions on product spaces. This consideration will naturally lead us to the well-known Fubini theorem.

If X and Y are locally compact Hausdorff spaces, so is $X \times Y$. Thus, we can define $C_0(X \times Y)$.

Theorem 5.   *Let X and Y be two locally compact Hausdorff spaces. Let I and J be two non-negative linear functionals on $C_0(X)$ and $C_0(Y)$, respectively. Then for each $f(x, y) \in C_0(X \times Y)$,*

$$K(f) = I_x(J_y f(x, y)) = J_y(I_x f(x, y))$$

is an integral on $C_0(X \times Y)$.

PROOF.   Let C be the support of f in $X \times Y$. Then $\bar{C}$ is compact and $C \subset C_1 \times C_2$, where C_1 and C_2 are compact in X and Y, respectively. By part (b) of the proof of Theorem 1, I and J are uniformly bounded on C_1 and C_2, respectively. Let M_1 and M_2 be their bounds. By the Stone-Weierstrass approximation theorem (§11, Chapter I) given $\varepsilon > 0$, we can find $f_i \in C_0(C_1)$ and $g_i \in C_0(C_2)$, $1 \leq i \leq n$, such that $\|f(x, y) - \sum_{i=1}^{n} f_i(x)g_i(y)\|_\infty < \varepsilon$. Whence it follows that

$$\left| J_y f(x, y) - \sum_{i=1}^{n} J_y f_i(x)g_i(y) \right| < \varepsilon M_2.$$

But then $J_y f(x, y)$, being the uniform limit of continuous functions $\sum_{i=1}^{n} J_y f_i(x) g_i(y)$, is a continuous function of x, and

$$\left| I_x J_y f - \sum_{i=1}^{n} I(f_i)J(g_i) \right| < \varepsilon M_1 M_2.$$

Similarly, now reversing the order, we get the following inequality:

$$\left| J_y I_x f - \sum_{i=1}^{n} I(f_i) J(g_i) \right| < \varepsilon M_1 M_2.$$

Hence, combining the two inequalities, we have

$$|I_x J_y f - J_y I_x f| < 2\varepsilon M_1 M_2.$$

Since ε is arbitrary, we have

$$I_x(J_y f(x, y)) = J_y(I_x f(x, y)).$$

Since K is non-negative and linear, K is an integral on $C_0(X \times Y)$ by Theorem 1.

The following theorem can be derived from the above (see Loomis,[29] p. 44 for a proof).

Corollary. (*Fubini theorem.*) *Let K be the functional defined on $C_0(X \times Y)$ as in Theorem 2. Let $f \geq 0$ be a Baire function on $X \times Y$. Then for each fixed $y, f(., y) \in \mathscr{B}^+(X), I_x(f(., y)) \in \mathscr{B}^+(Y)$ and*

$$K(f) = J_y (I_x(f(x, y))).$$

Usually, one studies the integration on the product space as follows: Let μ_1 and μ_2 be measures on locally compact spaces X and Y, respectively. For $A \in S(X)$ (Borel ring of X) and $B \in S(Y)$, define an elementary measure on "rectangles" $A \times B$ as follows:

(1) $\mu(A \times B) = \mu_1(A) \cdot \mu_2(B).$

Clearly, this is the same as starting with an integral

(2) $\displaystyle\int \sum_{i=1}^{n} c_i \chi_{A_i \times B_i} = \sum_{i=1}^{n} c_i \mu_1(A_i)\mu(B_i)$

on simple functions, i.e., finite linear conbinations of characteristic functions of sets in $S(X) \times S(Y)$. Thus, the whole theory of measures as developed via integrals can be shown to be equivalent to the usual theory starting with elementary measures. As the integral defined in (2) can be extended to a larger class of functions, so can the measure μ be extended to a regular measure on $X \times Y$.

35. EXISTENCE OF AN INVARIANT FUNCTIONAL

To establish a measure on a certain class of subsets it is sufficient to establish a linear functional with appropriate properties. This is precisely what we are going to do in the sequel for the existence of Haar measure.

Definition 2. (a) Let X be a set and R^X (or C^X) denote the set of all real- (or complex-) valued functions. $B(X)$ is the set of all bounded functions. It is easy to see that both R^X and $B(X)$ are linear spaces.

(b) Let G be a group and $f \in R^G$. Then for each fixed $a \in G$, f_a, defined by $f_a(x) = f(ax)$ for all $x \in G$, is called a *left translate* of f. Similarly f^a, defined by $f^a(x) = f(xa)$, a fixed and $x \in G$, is called a *right translate* of f.

(c) Let G be a group and $f \in R^G$. Then f is said to be *left* (or *right*) *invariant* if $f = f_a$ (or $f = f^a$).

(d) Let I be a linear functional on R^G. Then I is said to be *left-* (or *right-*) *invariant* if $I(f) = I(f_a)$ (or $I(f) = I(f^a)$).

(e) Let G be a topological group and $f \in R^G$. The set $\{x \in G : f(x) \neq 0\}$ is called the *support* of f.

(f) Let G be a Hausdorff topological group. The set $C_0(G)$, as usual, denotes the subset of R^G or C^G consisting of continuous functions having compact supports.

Since each continuous function on a compact set is bounded, it follows that $C_0(G) \subset C(G) \subset B(G)$, as mentioned before. For each $f \in B(G)$, define $\|f\| = \sup_{x \in G} |f(x)|$. Then it can be verified easily that $\|f\|$ is a norm and $B(G)$ is a Banach space. Also $C(G)$ is a closed normed subspace of $B(G)$ and, therefore, a Banach space. Furthermore, $C_0(G)$ is a normed subspace of $C(G)$. The closure of $C_0(G)$ in $C(G)$ consists of all continuous functions f vanishing at infinity on G, i.e., for any $\varepsilon > 0$ there exists a compact set K depending on f such that $|f(x)| < \varepsilon$ for all $x \notin K$. The set of functions vanishing at infinity will be denoted by $C_\infty(G)$.

(g) Let $C_0^+(G) = \{f \in C_0(G) : f \geq 0\}$.
Then for each $f \in C_0(G)$, $f^+, f^- \in C_0^+(G)$.

Our object is to prove the existence of an invariant nonzero linear functional on $C_0^+(G)$ and then extend the functional to the whole of $C_0(G)$ as follows:

Suppose I is such a linear functional on $C_0^+(G)$. Let $f \in C_0(G)$. Then $I(f^+)$ and $I(f^-)$ exist. Define the extension, again denoted by I, as:

$$I(f) = I(f^+) - I(f^-).$$

To prove the theorem concerning the existence of the required functional, we begin with the following preliminary propositions leading us to the required result. The ideas and proofs of the statement given here are due to Weil[46] and the pattern followed here is that of Loomis.[29]

Definition 3. Let f and g be any real-valued functions on a group G such that $g \geq 0$, $g \neq 0$. We consider all finite sequences of positive numbers c_i and all finite sets $\{x_i\} \subset G$ such that

$$f(x) \leq \sum_i c_i g(x_i^{-1} x)$$

for all $x \in G$. Define

$$(f:g) = \begin{cases} \inf \left\{ \sum_i c_i : f(x) \leq \sum_i c_i g(x_i^{-1}x) \right\} \\ \infty \quad \text{if no such systems of } \{c_i\} \\ \qquad \text{and } \{x_i\} \text{ exist.} \end{cases}$$

We gather some important properties of the function $(f:g)$ in the following:

> **Proposition 2.**   *Let f, g, h, f_1, $f_2 \in C_0^+(G)$ and let $g \neq 0$. Then:*
> (1) *$(f:g)$ exists and is finite;*
> (2) *for $f \neq 0$, $(f:g) > 0$;*
> (3) *$(f_a:g) = (f:g)$ for all $a \in G$;*
> (4) *$(cf:g) = c(f:g)$, $c \geq 0$;*
> (5) *$(f_1 + f_2:g) \leq (f_1:g) + (f_2:g)$;*
> (6) *$(f:g) \leq (f:h)(h:g)$;*
> (7) *$f_1 \leq f_2$ implies $(f_1:g) \leq (f_1:g)$.*

Proof.   (1) Let C be the compact support of f. Since $g \neq 0$, there exists an open neighborhood U of the identity $e \in G$ and a point $a \in G$ such that $g(y) \geq m > 0$ for all $y \in aU$, where m is the infimum of g on aU. Since C is compact, there is a finite number of sets $\{y_i U\}(1 \leq i \leq n)$ that cover C. Hence, it follows that for all $x \in G$,

$$f(x) \leq \frac{\sup f}{m} \sum_{i=1}^{n} g(ay_i^{-1}x).$$

Whence we see that $(f:g)$ exists and

$$(f:g) \leq \frac{n \sup f}{m} < \infty.$$

(2) Since f, $g \neq 0$, $\sup f > 0$ and $\sup g > 0$. Further, since $g \in C_0^+(G)$, g is bounded; therefore, $0 < \sup g < \infty$. Now the inequality

$$f(x) \leq \sum_i c_i g(x_i^{-1}x)$$

for all $x \in G$ implies

$$(f:g) \geq \frac{\sup f}{\sup g} > 0.$$

(3) Since $f_a(x) = f(ax)$,

$$f(x) \leq \sum_i c_i g(x_i^{-1}x), \qquad \text{for all } x \in G$$

implies

$$f_a(x) = f(ax) \leq \sum_i c_i g(x_i^{-1}ax) = \sum_i c_i g((a^{-1}x_i)^{-1}x).$$

Thus $(f_a:g) \leq (f:g)$ for all a in G and hence (3) follows.

(4) For $c \geq 0$,

$$cf(x) \leq \sum_i cc_i g(x_i^{-1}x).$$

Hence

$$(cf:g) = \inf \sum_i cc_i = c \inf \sum_i c_i = c(f:g).$$

(5) Let

$$f_1(x) \leq \sum_j a_j g(x_j^{-1}x)$$

and

$$f_2(x) \leq \sum_k b_k g(y_k^{-1}x).$$

Then

$$f_1(x) + f_2(x) \leq \sum_j a_j g(x_j^{-1}x) + \sum_k b_k g(y_k^{-1}x).$$

Taking infima successively in $\{a_j\}$ and $\{b_k\}$ we find

$$(f_1 + f_2 : g) \leq (f_1 : g) + (f_2 : g).$$

(6) If

$$f(x) \leq \sum_i a_i h(x_i^{-1}x)$$

and

$$h(x) \leq \sum_j b_j g(y_j^{-1}x),$$

then

$$f(x) \leq \sum_{i,j} a_i b_j g(y_j^{-1}x_i^{-1}x).$$

Hence

$$(f:g) \leq \sum_{i,j} a_i b_j$$

$$\leq \left(\sum_i a_i\right)\left(\sum_j b_j\right).$$

Taking infima successively, we have

$$(f:g) \leq (f:h)(h:g).$$

(7) The proof is clear.

Definition 4. Let $f_0 \in C_0^+(G)$, $f_0 \neq 0$ and be fixed once and for all. Define

$$I_g(f) = \frac{(f:g)}{(f_0:g)}.$$

The properties of $I_g(f)$ corresponding to those of $(f:g)$ in Proposition 2 can also be established, viz.:

Proposition 3. *Let f, g, f_1, f_2 be the same as in Proposition 2. Then:*
(1') $I_g(f) \geq 0$, *and* $I_g(f) = 0$ *if, and only if,* $f = 0$;
(2') $I_g(f_a) = I_g(f)$;

(3') $I_g(cf) = cI_g(f), c \geq 0$;

(4') $I_g(f_1 + f_2) \leq I_g(f_1) + I_g(f_2)$;

(5') $\dfrac{1}{(f_0:f)} \leq I_g(f) \leq (f:f_0), f \neq 0$;

(6') $f_1 \leq f_2 \Rightarrow I_g(f_1) \leq I_g(f_2)$.

PROOF. (1') By definition clearly $I_g(f) \geq 0$. Now suppose $f = 0$. Then $(f:g) = 0$, as is easy to see. Since $f_0 \neq 0$, $(f_0:g) > 0$ by (2) of Proposition 2 and hence

$$I_g(f) = \frac{(f:g)}{(f_0:g)} = 0.$$

Conversely, suppose $I_g(f) = 0$. If $f \neq 0$, then by (2) of Proposition 2, $(f:g) > 0$ and also $(f_0:g) > 0$. Hence, $I_g(f) > 0$, which is a contradiction. Therefore $f = 0$.

(2'), (3'), and (4') are immediate consequences of (3), (4), and (5) of Proposition 2, respectively.

(5') From (6) of Proposition 2 we have

$$(f:g) \leq (f:f_0)(f_0:g)$$

and

$$(f_0:g) \leq (f_0:f)(f:g).$$

Hence

$$\frac{1}{(f_0:f)} \leq \frac{(f:g)}{(f_0:g)} \leq (f:f_0),$$

whence (5').

(6') The result is an immediate consequence of (7) of Proposition 2.

Proposition 4. *Let $f_1, f_2 \in C_0^+(G)$, where G is a locally compact Hausdorff group. Let $\varepsilon > 0$ be given. Then there exists a neighborhood U of $e \in G$ such that for all $g \in C_0^+(G)$ having supports in U,*

$$I_g(f_1) + I_g(f_2) \leq I_g(f_1 + f_2) + \varepsilon.$$

PROOF.   Since $f_1, f_2 \in C_0^+(G)$, $f_1 + f_2 \in C_0^+(G)$. Let C be the support of $f_1 + f_2$. Then there exists a function $f' \in C_0^+(G)$ such that $f'(C) = 1$ by Corollary 1, §26, Chapter IV, since G is locally compact.

Let δ, ε' be arbitrary positive numbers and let $f = f_1 + f_2 + \delta f'$. Clearly $f \in C_0^+(G)$. Let

$$h_i(x) = \begin{cases} \dfrac{f_i(x)}{f(x)} & \text{if } f(x) \neq 0 \\ 0 & \text{otherwise} \end{cases}$$

for $i = 1, 2$. Then

$$h_1 + h_2 = \frac{f_1 + f_2}{f} \leq 1,$$

and $h_i \in C_0^+(G)$. Hence, by Proposition 5, §30, Chapter V, there exists an open neighborhood U of $e \in G$ such that for $x, y \in G$,

$$|h_i(x) - h_i(y)| < \varepsilon',$$

whenever $x^{-1}y \in U$, $i = 1, 2$.

Let g be a function in $C_0^+(G)$ with its support in U. Consider

$$f(x) \leq \sum_j c_j g(x_j^{-1}x).$$

If $x_j^{-1}x$ is such that $g(x_j^{-1}x) \neq 0$, then $x_j^{-1}x \in U$ and, therefore,

$$|h_i(x) - h_i(x_j)| < \varepsilon',$$

or

$$h_i(x_j) - \varepsilon' < h_i(x) < h_i(x_j) + \varepsilon'.$$

But then,

$$\begin{aligned} f_i(x) &= f(x)h_i(x) \\ &\leq \sum_j c_j g(x_j^{-1}x)h_i(x) \\ &\leq \sum_j c_j g(x_j^{-1}x)(h_i(x_j) + \varepsilon'). \end{aligned}$$

Hence,

$$(f_i : g) \leq \sum_j c_j (h_i(x_j) + \varepsilon'),$$

and so

$$(f_1 : g) + (f_2 : g) \leq \sum_j c_j [h_1(x_j) + h_2(x_j) + 2\varepsilon']$$

$$\leq (1 + 2\varepsilon') \sum_j c_j.$$

This shows that

$$(f_1 : g) + (f_2 : g) \leq (f : g)(1 + 2\varepsilon').$$

Let $f_0 \in C_0^+(G), f_0 \neq 0$ be fixed. Dividing the above inequality by $(f_0 : g)$, we have

$$I_g(f_1) + I_g(f_2) \leq I_g(f)(1 + 2\varepsilon').$$

Since $f = f_1 + f_2 + \delta f'$, by (3') and (4') of Proposition 3 we have

$$I_g(f) \leq I_g(f_1) + I_g(f_2) + \delta I_g(f').$$

Hence,

$$I_g(f_1) + I_g(f_2) \leq [I_g(f_1) + I_g(f_2) + \delta I_g(f')](1 + 2\varepsilon')$$

$$\leq I_g(f_1) + I_g(f_2) + \delta I_g(f') + 2\varepsilon'[I_g(f_1) + I_g(f_2) + \delta I_g(f')].$$

Since $I_g(f_1), I_g(f_2)$, and $I_g(f')$ are finite and ε', δ are arbitrary, we can choose ε' and δ such that

$$\{\delta I_g(f') + 2\varepsilon'[I_g(f_1) + I_g(f_2) + \delta I_g(f')]\} < \varepsilon.$$

This completes the proof of the proposition.

Now we prove the existence of an invariant functional on $C_0^+(G)$.

Theorem 6. *Let G be a locally compact Hausdorff topological group. Then there exists a nontrivial (i.e., not identically zero), non-negative, left-invariant, positive homogeneous, additive, functional I on $C_0^+(G)$.*

PROOF. For each nonzero $f \in C_0^+(G)$, let X_f denote the closed bounded interval $\left[\dfrac{1}{(f_0:f)}, (f:f_0)\right]$. By Tychonoff's theorem, $X = \prod_f X_f$ is a compact Hausdorff space. For each $g \neq 0$, $I_g = (I_g(f)) \in X$, since $I_g(f) \in X_f$ by Proposition 3.

For each neighborhood V of $e \in G$, let

$$F_V = Cl\{I_g \in X : \text{support of } g \text{ is in } V\},$$

where Cl denotes the closure. The family $\{F_V\}$ of closed sets in X, when V runs over the family of neighborhoods of e, has the finite intersection property. For, if $F_{V_1}, \ldots, F_{V_n}$ is a finite subfamily, then for any g with its support in $V = \bigcap_{1=1}^{n} V_i$, $I_g \in \bigcap_{i=1}^{n} F_{V_i}$. Since X is compact, $\bigcap F_V \neq \emptyset$ and, hence, there exists I such that $I \in F_V$ for each neighborhood V of $e \in G$. Thus, $I = (I(f))$, where $I(f)$ is the fth coordinate of I. We show that $I(f)$ is the required linear functional.

Since $I \in F_V$ for each V, for $\varepsilon > 0$ and any finite family of $f_i \in C_0^+(G)$, $(1 \leq i \leq n)$, there exists a g with support in V for a given V such that

(A) $|I(f_i) - I_g(f_i)| < \varepsilon, 1 \leq i \leq n.$

Since (i) to (iv) below hold for $J = I_g$, (i) to (iv) obtain for $J = I$. For $f, f_1, f_2 \in C_0^+(G)$:

(i) $J \neq 0, J(f) > 0$ if $f \neq 0$,
(ii) $J(f_a) = J(f)$,
(iii) $J(cf) = cJ(f), c \geq 0$,
(iv) $J(f_1 + f_2) \leq J(f_1) + J(f_2)$.

Thus it follows that I is a nontrivial, left-invariant, non-negative, positive homogeneous and subadditive real-valued function. To complete the proof we must show that I is additive. In other words, we must show that the reverse inequality in (iv) holds for $J = I$.

For a given $\varepsilon > 0$, by the above approximation inequality (A), it follows that

$$|I(f_1 + f_2) - I_g(f_1 + f_2)| < \varepsilon.$$

By Proposition 4,

$$I_g(f_1) + I_g(f_2) \leq I_g(f_1 + f_2) + \varepsilon.$$

Therefore,

$$I(f_1) + I(f_2) \leq I(f_1 + f_2) + 4\varepsilon.$$

Since ε is arbitrary, we have

(v) $I(f_1) + I(f_2) \leq I(f_1 + f_2).$

Combining (iv) and (v), we see that $J = I$ is additive.

36. ESSENTIAL UNIQUENESS OF THE HAAR INTEGRAL

In this section we show that the additive functional I, the existence of which has been shown in the previous section, is "essentially" unique. More precisely, we have the following:

Theorem 7. *Let G be a locally compact Hausdorff topological group. Then the invariant functional I, determined in Theorem 6, is "essentially" unique, i.e., if there is another such function J on $C_0^+(G)$ then there exists a real number $c > 0$ such that*

$$J(f) = cI(f)$$

for all $f \in C_0^+(G)$.

PROOF. Suppose there exists another invariant functional J defined on $C_0^+(G)$ and having the same properties as those of I.

Let $f \in C_0^+(G)$, $f \neq 0$, and let the support of f be C. Since G is locally compact, there exists an open set U such that $C \subset U$ and $\bar{U}$ is compact. Furthermore, by Proposition 5, §30, Chapter V, f is left uniformly continuous because $f \in C_0^+(G)$. Therefore, for $\varepsilon > 0$ there exists a symmetric neighborhood V of e such that for all $x, y \in G$, $y^{-1}x \in V$ implies

    (i) $|f(yx) - f(x)| < \varepsilon/2$

    (ii) $|f(xy) - f(x)| < \varepsilon/2$

and (iii) $VC \cup CV \subset U$.

Finally, let $f' \in C_0^+(G)$ such that $f'(U) = 1$. Such an f' exists because G, being a Hausdorff locally compact group, is normal and therefore Urysohn's lemma applies.

From (iii) for $y \in V$, we have $f(xy) = f(xy)f'(x)$ and $f(yx) = f(yx)f'(x)$, because $f'(U) = 1$ and $f(G \sim U) \subset f(G \sim C) = 0$.

Thus from (i) and (ii) we have

$$|f(xy) - f(yx)| < \varepsilon f'(x)$$

for all $x \in G$ and $y \in V$.

Let h be a positive function in $C_0^+(G)$ such that the support of h lies in V and $h(x) = h(x^{-1})$. Then

$$
\begin{aligned}
I(h) J(f) &= I_y(h(y)) J_x(f(x)) \\
&= I_y(h(y)) J_x f(yx) && \text{(owing to left-invariance of J)} \\
&= I_y J_x(h(y) f(yx)) && \text{(by the Fubini theorem, §33).}
\end{aligned}
$$

Furthermore,

$$
\begin{aligned}
J(h) I(f) &= J_x(h(x)) I_y(f(y)) \\
&= J_x(h(y^{-1}x)) I_y(f(y)) && \text{(by left-invariance of J)} \\
&= J_x I_y(h(y^{-1}x) f(y)) && \text{(by the Fubini theorem, §33)} \\
&= J_x I_y(h(xy^{-1}) f(y)) && \text{(since h is symmetric)} \\
&= J_x I_y(h(y) f(xy)) \\
&= I_y J_x(h(y) f(xy)).
\end{aligned}
$$

Hence

(A)
$$|I(h)J(f) - J(h)I(f)| \leq I_y J_x |f(yx) - f(xy)| \, h(y)$$
$$\leq \varepsilon I_y J_x f'(x) h(y)$$
$$\leq \varepsilon I(h) J(f').$$

Similarly, if $g \in C_0^+(G)$ and if g has properties similar to those of h in the above calculations, then we have

(B)
$$|I(h) J(g) - I(h) I(g)| \leq \varepsilon I(h) J(g')$$

in which g' has the same relation to g as f' to f. Thus, by dividing (A) and (B) by $I(f)I(h)$ and $I(g)I(h)$, respectively, we have

$$\left| \frac{J(f)}{I(f)} - \frac{J(h)}{I(h)} \right| \leq \varepsilon \frac{J(f')}{I(f)}$$

and

$$\left| \frac{J(g)}{I(g)} - \frac{J(h)}{I(h)} \right| \leq \varepsilon \frac{J(g')}{I(g)}.$$

The latter two inequalities imply

$$\left| \frac{J(f)}{I(f)} - \frac{J(g)}{I(g)} \right| \leq \varepsilon \left| \frac{J(f')}{I(f)} + \frac{J(g')}{I(g)} \right|.$$

Since ε is arbitrary, we have

$$\frac{J(f)}{I(f)} = \frac{J(g)}{I(g)}.$$

If $g \neq 0$ is fixed, then $\dfrac{J(g)}{I(g)} = c > 0$ and, therefore,

$$J(f) = cI(f)$$

for all $f \in C_0^+(G)$. This completes the proof.

Remark. If G is abelian, the proof of Theorem 7 is very simple and is left for the reader to verify. Moreover, in that case, every left-invariant integral is also a right-invariant integral.

Definition 5. The measure μ induced by I is called *Haar measure.* The linear functional I is called *Haar integral.*

For each compact set $A \subset G$,

$$\mu(A) = \inf \{I(f) : f \geq 1 \text{ on } A, f \in C_0(G)\}.$$

For each nonempty open set B of G, $\mu(B) > 0$.

Proposition 5. *A locally compact Hausdorff topological group G is compact if, and only if, $\mu(G) < \infty$.*

PROOF. Suppose G is compact. Then the constant functions are I-summable. Therefore, the characteristic function $\chi_G \in C_0(G)$ and, hence, $\mu(G) = I(\chi_G) < \infty$.

Conversely, suppose $\mu(G) < \infty$ and that G is not compact. Let V be a compact neighborhood of e. Then G cannot be covered by a finite number of translates of V. Therefore, there exists a sequence $\{x_n\}$ such that $x_n \notin \bigcup\limits_{i=1}^{n-1} x_i V$. Let U be a symmetric open neighborhood of $e \in G$ such that $U^2 \subset V$. Then $\{x_n U\}$ is a sequence of pairwise disjoint open sets of G. For, if $x_m U \cap x_n U \neq \varnothing$, $m > n$, then $x_m \in x_n U^2 \subset x_n V$ is a contradiction. Since $\mu(x_n U) = a > 0$ for each $n \geq 1$, we see $\mu(G) \geq \sum\limits_{n=1}^{\infty} \mu(x_n U) = \infty$, contrary to the assumption. This proves the proposition.

We shall see in the sequel that a left-invariant Haar measure is not necessarily right-invariant.

Definition 6. (a) A locally compact topological group is said to be *unimodular*, if left-invariant measure coincides with right-invariant measure (up to a multiplicative constant).

Let I be a left-invariant integral on a locally compact Hausdorff topological group. Then $I(f_a) = I(f)$ for each $f \in C_0(G)$. Consider $I(f^x)$, which is the left-invariant Haar integral for a fixed $x \in G$. Therefore, by the uniqueness theorem, there exists a positive real number $\Delta(x)$, depending on x, such that

$$I(f^x) = \Delta(x) \, I(f).$$

(b) The function $\Delta(x)$ is called the *modular* function.

Proposition 6. *A locally compact Hausdorff topological group is unimodular if, and only if, $\Delta(x) = 1$ for all $x \in G$.*

PROOF. Obvious.

Theorem 8. *A compact Hausdorff topological group G is unimodular.*

PROOF. Since $\chi_G \in C_0(G)$, we have $I(\chi_G^x) = \Delta(x) I(\chi_G) = \Delta(x)\mu(G)$. But $I(\chi_G^x) = \mu(G)$. Hence, $\mu(G) = \Delta(x)\mu(G)$. By Proposition 5, $\mu(G) < \infty$ and so $\Delta(x) = 1$ for all $x \in G$.

Theorem 9. *A locally compact Hausdorff topological abelian group is unimodular.*

PROOF. Trivial.

Proposition 7. $\Delta(x)$ *is a continuous homomorphism of the locally compact group G into the set of positive real numbers.*

PROOF. Clearly Δ is a positive-valued function. For $x, y \in G$ and $f \in C_0(G)$,

$$I(f_{xy}) = \Delta(xy) \, I(f).$$

But if we compute $\Delta(xy)$ successively, then we have

$$\Delta(xy)I(f) = I(f_{xy}) = \Delta(x)I(f_y) = \Delta(x)\Delta(y)I(f).$$

Since $I(f) < \infty$,

$$\Delta(xy) = \Delta(x)\Delta(y).$$

That means Δ is a homomorphism of G into the multiplicative group of positive real numbers.

Now for continuity of Δ it is sufficient to show that $\Delta(x)$ is continuous at the unity $e \in G$, because Δ is a homomorphism. Let U be a neighborhood of $e \in G$ such that $\bar{U}$ is compact. Let $f \in C_0^+(G), f \neq 0$. Let $A = \{x \in G : f(x) > 0\}$. Then $\bar{A}$ is compact because it is a closed subset of the compact support of f. But then $B = \bar{A}\,\bar{U}$ is also compact. There exists a $g \in C_0^+(G)$ such that $g(B) = 1$. Let $\varepsilon > 0$ be given. By Proposition 5, §30, Chapter V, there exists a symmetric neighborhood V of e such that $V \subset U$ and whenever $x^{-1}y \in V$ for $x, y \in G$,

$$|f(x) - f(y)| \leq \varepsilon \frac{I(f)}{I(g)}.$$

Thus, if $x \in V$, we have

$$|f(yx^{-1}) - f(y)| \leq \varepsilon \frac{I(f)}{I(g)} g(y)$$

for all $y \in G$. Hence, we have

$$|I(f^{x^{-1}}) - I(f)| \leq \varepsilon \frac{I(f)}{I(g)} I(g) = \varepsilon I(f).$$

This proves that

$$|\Delta(x) - 1| = \left| \frac{I(f^{x^{-1}})}{I(f)} - 1 \right| \leq \varepsilon$$

for all $x \in V$. Therefore, $\Delta(x)$ is continuous.

37. COMPUTATION OF HAAR INTEGRALS IN SPECIAL CASES

We shall assume that the reader is familiar with the elementary theory of functions of several variables, in particular, Riemann integration and Jacobians of transformations in an n-dimensional Euclidean space R^n.

Let X be an open subset of R^n such that

(i) Relative to some definition of product, X is a topological group. If $x, y \in X$, xy denotes the group product of x and y. (Note that if $f:(x, y) \to xy$ then f maps $X \times X \subset R^{n^2}$ into $X \subset R^n$. We can write f as $f(x, y) = f(x_1, \ldots, x_n; y_1, \ldots, y_n) \in R^n$.)

(ii) For each i, $f_i = p_i \circ f$ is continuously differentiable, where p_i is the ith projection of R^n.

(iii) The *Jacobians* $J(l_a)$ *and* $J(r_a)$ of the transformations l_a and r_a, respectively, are constant. (Such transformations exist, e.g., all linear transformations of R^{n^2} into R^{n^2} have constant Jacobians. In our case, if $f_i = \sum_{j,k} c_{jk}^{(i)} x_j x_k + a_j$, $c_{jk}^{(i)}$, a_j are all real, then the conditions are satisfied.)

It is clear that the translations have the following properties, viz., $l_a \circ l_b = l_{ab}$, $r_a \circ r_b = r_{ba}$ for $a, b \in X$. If l_e is the identity transformation then $|J(l_e)| = 1$. Similarly, $|J(r_e)| = 1$. Recall that $J(l_a \circ l_b) = J(l_a)J(l_b)$. Hence, $a \to |J(l_a)|$ is a continuous mapping of X into the multiplicative group R^+ of positive real numbers and is a homomorphism of X into R^+.

Let f be a continuous complex-valued function defined on X and vanishing outside a compact set. As usual, the symbol

$$\int \cdots \int f(x_1, \ldots, x_n) \, dx_1 \, dx_2 \cdots dx_n \equiv \int_X f(x) \, dx$$

denotes the ordinary Riemann integral of f, in which $x = (x_1, \ldots, x_n)$.

Proposition 8. *The functional* $I(f) = \int_X f(x) \, |J(l_x)|^{-1} \, dx$ *is a left Haar integral. Similarly,*

$$I'(f) = \int_X f(x) \, |J(r_x)|^{-1} \, dx$$

is a right Haar integral.

PROOF. If $f \in C_0^+(X)$, then clearly $I(f) \geq 0$. The fact that the Riemann integral is linear implies the linearity of I as well. Thus, the only property left to be shown is the left invariance of the first integral. The right invariance of I' can be established similarly. First we recall that $l_a(X) = X$ and that for any $g \in C_0(X)$,

$$\int_{l_a(X)} g(x) \, dx = \int_X g(l_a(y)) \, |J(l_a(y))| \, dy,$$

as is well known from the theory of functions of several variables. Thus, if we put $g = f_{a^{-1}} |J(l_x)|^{-1}$ and $ax = y$ in the above formula, we get

$$I(f_{a^{-1}}) = \int_X f_{a^{-1}}(x) \, |J(l_x(x))|^{-1} \, dx$$

$$= \int_X f_{a^{-1}}(l_a(y)) \, |J(l_{ay}(y))|^{-1} \, |J(l_a(y))| \, dy$$

$$= \int_X f(l_{a^{-1}}(l_a(y))) \, |J(l_a)|^{-1} |J(l_y)|^{-1} \, |J(l_a)| \, dy$$

$$= \int_X f(y) \, |J(l_y)|^{-1} dy = I(f).$$

This proves the left invariance of I, and thus the proposition is established.

The following are particular cases of the above:

Example 1. The functional $I(f) = \int_{-\infty}^{\infty} \dfrac{f(x)}{|x|} \, dx$ is a left and right Haar integral on the multiplicative group $R \sim \{0\}$ of the nonzero real numbers, in which $f \in C_0(R \sim \{0\})$ and the integral is extended over $[-b, -a] \cup [a, b]$, where a, b are positive real numbers such that $a < b$. (The Jacobian of the transformation: $x \to ax$ is equal to a.)

Example 2. The functional

$$I(f) = \int_{-\infty}^{\infty} \int_{-\infty}^{\infty} \frac{f(z)}{|z|^2} \, dx, \, dy = \int_{-\infty}^{\infty} \int_{-\infty}^{\infty} \frac{f(x + iy)}{x^2 + y^2} \, dx, \, dy,$$

$z = x + iy$, $i = \sqrt{-1}$, is a Haar integral on the multiplicative group $C \sim \{0\}$ of nonzero complex numbers, in which, as usual, $f \in C_0(C \sim \{0\})$ and the double integral has a meaning similar to that described in Example 1. (The Jacobian of the transformation: $z \to az$ is equal to $a_1^2 + a_2^2$, where $a = a_1 + ia_2$.)

Example 3. Let G be the set of all matrices

$$X = \begin{bmatrix} x & y \\ 0 & 1 \end{bmatrix},$$

where $x > 0$ and y runs over the set of all real numbers. Under matrix multiplication G is a group with

$$\begin{bmatrix} 1 & 0 \\ 0 & 1 \end{bmatrix}$$

as its unit element. The group G can be identified with the right half-plane of R^2. Clearly G is a locally compact Hausdorff topological group and satisfies all the conditions (i) to (iii) considered in the earlier paragraphs of this section. For each fixed element

$$A = \begin{bmatrix} a & b \\ 0 & 1 \end{bmatrix},$$

$A \in G$, left translation by A is given by

$$AX = \begin{bmatrix} a & b \\ 0 & 1 \end{bmatrix}\begin{bmatrix} x & y \\ 0 & 1 \end{bmatrix} = \begin{bmatrix} ax & ay + b \\ 0 & 1 \end{bmatrix},$$

and right translation by A is given by

$$XA = \begin{bmatrix} x & y \\ 0 & 1 \end{bmatrix}\begin{bmatrix} a & b \\ 0 & 1 \end{bmatrix} = \begin{bmatrix} ax & bx + y \\ 0 & 1 \end{bmatrix}.$$

Therefore, the Jacobians of the left and right translations are, respectively, a^2 and a. Hence, the left and right Haar integrals, respectively, are:

$$\int_{-\infty}^{\infty} \int_{0}^{\infty} \frac{f(x, y)}{x^2} \, dx \, dy$$

and

$$\int_{-\infty}^{\infty} \int_{0}^{\infty} \frac{f(x, y)}{|x|} \, dx \, dy,$$

for any $f \in C_0(G)$, and as usual the integrals are extended over a bounded closed subset of the open right half-plane.

This provides a well-known example of a locally compact group, the left and right Haar integrals of which do not have a constant ratio.

Exercises

1. (a) *Hölder's inequality:* If $f \in L_p$ and $g \in L_q$, $(p^{-1} + q^{-1} = 1)$, then $fg \in L_1$ and $|fg| \le \| \, \|_p \, \|g\|_q$.
   (b) *Minkowski's inequality:* If $f, g \in L_p$ $(p \ge 1)$ then $f + g \in L_p$.
   (c) $L_p (p \ge 1)$ is a Banach space and L_2 a Hilbert space.
2. Let G be a topological group. Then for any $f \in C(G)$, $g(x) = f(x) + f(x^{-1})$, $x \in G$, is a symmetric function in $C(G)$.
3. Find Haar measure on the following groups: (a) Discrete groups. Show that a discrete group is unimodular; (b) Quotient groups G/H, where H is a closed invariant subgroup of a locally compact Hausdorff topological group G; (c) Product groups $G = \prod_{\alpha} G_\alpha$, where $\{G_\alpha\}$ is a family of compact Hausdorff topological groups G_α.
4. Let G be a locally compact Hausdorff topological group.
   (a) For $f \in L_1(G)$, show that

$$\int_G f(x^{-1}) \, \Delta(x^{-1}) \, dx = \int_G f(x) \, dx;$$

   (b) From (a), derive that Haar measure μ is inverse invariant (i.e., $\mu(E^{-1}) = \mu(E)$) if, and only if, G is unimodular;
   (c) Put $f^*(x) = \overline{f(x^{-1})} \, \Delta(x^{-1})$. Show that $f^* \in L_1(G)$ if, and only if, $f \in L_1(G)$.

Appendix to Chapter VI

Cartan's Proof of Existence and Uniqueness of the Haar Integral

Remark. Observe that in Weil's proof of the existence of Haar integral, the axiom of choice is used via Tychonoff's theorem. Moreover, the "essential" uniqueness is established separately. There is an elegant proof due to Cartan,[8] of the existence of Haar integral in which the Tychonoff theorem is avoided and the "essential" uniqueness is established simultaneously. We give here Cartan's proof for comparison.

Lemma 1 (*Partition of Unity*). *Let $\{C_i\}$ $(1 \leq i \leq n)$ be a finite family of compact subsets of a locally compact Hausdorff topological group G. Then there exists a finite family $\{f_i\}$ $(1 \leq i \leq n)$ of real-valued continuous functions on G such that $\sum_{i=1}^{n} f_i = 1$ and $f_i = 0$ on $G \sim C_i$ for each i.*

Use Corollary 1, §26, Chapter IV, and apply induction.

Lemma 2. *Let $f_i \in C_0^+(G)$, $1 \leq i \leq n$, and let $\varepsilon > 0$, $\delta > 0$ be given. There exists a neighborhood U of the identity $e \in G$ such that for all $g \in C_0^+(G)$ with support in U,*

$$I_g\left(\sum_{i=1}^{n} \lambda_i f_i\right) \leq \sum_{i=1}^{n} \lambda_i I_g(f_i) \leq I_g\left(\sum_{i=1}^{n} \lambda_i f_i\right) + \varepsilon$$

for all $0 \leq \lambda_i \leq \delta$.

This follows by the repeated use of Propositions 3 and 4, §35.

Theorem (*Approximation*). *Let $f \in C_0^+(G)$, $\varepsilon > 0$ and let V be a neighborhood of $e \in G$ such that $y^{-1}x \in V$ implies $|f(x) - f(y)| \leq \varepsilon$ for all $x, y \in G$. Let $h \in C_0^+(G)$, $h \neq 0$, be such that $h(G \sim V) = 0$. Then for each $\varepsilon' > \varepsilon$ one can find a finite family $\{s_i\} \subset G$ and real numbers $c_i > 0$ such that for all $x \in G$,*

$$\left| f(x) - \sum_{i \leq n} c_i h(s_i^{-1} x) \right| < \varepsilon'.$$

PROOF.   By hypothesis.

(1) $[f(x) - \varepsilon]h(s^{-1}x) \leq f(s)h(s^{-1}x) \leq [f(x) + \varepsilon]h(s^{-1}x).$

For a given $\varepsilon' > \varepsilon$, choose $\delta' > 0$ such that $(f{:}h^*)\delta' < \varepsilon' - \varepsilon$, in which $h^*(x) = h(x^{-1})$. Let W be a neighborhood of the identity such that $y^{-1}x \in W$, $(x, y \in G)$, implies $|h(x) - h(y)| \leq \delta'$. Since the support of f is compact, there exists a finite family $\{s_i\} \subset G$ $(1 \leq i \leq n)$ such that $\{s_iW\}$ covers the support of f. By Lemma 1, there exists $h_i \in C_0^+(G)$ such that $\sum_{i=1}^{n} h_i = 1$ for all the points of the support of f, and $h_i(G \sim s_iW) = 0$ for each i. Thus we have

$$h_i(s)f(s)[h(s_i^{-1}x) - \delta'] \leq h_i(s)f(s)h(s^{-1}x) \leq h_i(s)f(s)[h(s_i^{-1}x) + \delta']$$

for each i. Summing over i, we have

(2) $\sum_{i=1}^{n} h_i(s)f(s)h(s_i^{-1}x) - \delta'f(s) \leq f(s)h(s^{-1}x)$

$$\leq \sum_{i=1}^{n} h_i(s)f(s)h(s_i^{-1}x) + \delta'f(s).$$

From (1) and (2) we obtain

$$[f(x) - \varepsilon]h(s^{-1}x) - \delta'f(s) \leq \sum_{i=1}^{n} h_i(s)f(s)h(s_i^{-1}x)$$

$$\leq [f(x) + \varepsilon]h(s^{-1}x) + \delta'f(s).$$

Hence, for any $g \in C_0^+(G)$,

$$[f(x) - \varepsilon]I_g(h(s^{-1}x)) - \delta'I_g(f) \leq I_g\left[\sum_{i=1}^{n} h_i(s)f(s)h(s_i^{-1}x)\right]$$

$$\leq [f(x) + \varepsilon]I_g(h^*) + \delta'I_g(f).$$

[Observe that I_g is applied here on the function of s — not x — and also $I_g(h(s^{-1}x)) = I_g(h^*(x^{-1}s)) = I_g(h^*).$]
However,

$$I_g(f)/I_g(h^*) \leq (f{:}h^*) = \frac{\beta - \varepsilon}{\delta'},$$

for some $\beta < \varepsilon'$. Thus we have

(3) $f(x) - \beta \leq I_g\left(\sum_i \frac{h(s_i^{-1}x)}{I_g(h^*)}\, h_i(s_i^{-1}x)f(s)\right) \leq f(x) + \beta.$

Now putting $f_i = h_if$, $\lambda_i = h(s_i^{-1}x)/I_g(h^*)$, and $\delta = (f_0{:}h^*)\sup g$, by Lemma 2 there exists a neighborhood U of $e \in G$ such that

$$I_g\left(\sum_i \frac{h(s_i^{-1}x)}{I_g(h^*)}\, h_if\right) \leq \sum \frac{I_g(h_if)}{I_g(h^*)}\, h(s_i^{-1}x) \leq I_g\left(\sum \frac{h(s_i^{-1}x)}{I_g(h^*)}\, h_if\right) + \rho,$$

in which $\rho = \varepsilon' - \beta$. Now putting

$$c_i = I_g(h_if)/I_g(h^*),$$

from (3) we find that

$$f(x) - \beta \le \sum_i c_i h(s_i^{-1}x) \le f(x) + \rho + \beta.$$

Thus we have

$$f(x) - \varepsilon' \le \sum_i c_i h(s_i^{-1}x) \le f(x) + \varepsilon',$$

and the theorem is established.

Now from this theorem, one can derive the existence of an invariant functional $I(f)$ with the desired properties.

Observe that we can regard $I_g(f)$ as a function of g. For a given $f \in C_0^+(G)$ and $\varepsilon > 0$, we establish the existence of a neighborhood U of $e \in G$ such that if $g, k \in C_0^+(G)$ and if g and k have support in U, then

(4) $|I_g(f) - I_k(f)| \le \varepsilon.$

For this, let $\varepsilon' > 0$ be given. By the approximation theorem, there exists a function $h \in C_0^+(G)$, $s_i \in G$ $(1 \le i \le n)$, and $c_i > 0$ such that

$$\left| f(x) - \sum_i c_i h(s_i^{-1}x) \right| \le \varepsilon',$$

in which $\varepsilon' > \varepsilon$. By Lemma 2, there is a neighborhood U of $e \in G$ such that for any $g \in C_0^+(G)$ with its support in U,

$$I_g(f) - \varepsilon' \le \left(\sum c_i \right) I_g(f) \le I_g(f) + \varepsilon'.$$

Similarly, one has a corresponding inequality for $f_0 \in C_0^+(G)$, in which $I_g(f_0) = 1$ and g is the same as that for f. Now, combining with a similar inequality for $k \in C_0^+(G)$ with its support in U, we can establish (4).

But then the existence of a limit for $I_g(f)$ as g runs over the filter-base $\{g \in C_0^+(G):$ support of g in $U\}$ is immediately assured. Also this limit $I(f)$ is additive in f and has the desired properties of an invariant functional. Moreover, I is unique up to a positive factor. All these facts can be established as in Weil's proof.

VII

Finite-Dimensional
Representations
of
Groups

In this chapter we shall discuss continuous homomorphisms of a topological group into various groups of matrices. The device of mapping a topological group into a group of matrices transforms the study of abstract groups into that of "concrete ones." It should be observed that groups of matrices can be replaced by groups of transformations on a Hilbert space and, thus, the study of representations can be made in a more general setting. (This program has been discussed in exercises at the end of this chapter.) In this chapter, however, we confine ourselves to matrices. After establishing Schur's lemma and the orthogonality relations, we shall show that, for a metrizable compact group, the set of all orthonormal continuous complex-valued functions is countable. Via the famous theorem of Peter-Weyl, we shall show that a compact metrizable topological group is homeomorphic with the inverse limit of a sequence of Lie groups.

38. SCHUR'S LEMMA

We begin with a few remarks about matrices. As mentioned in §27 after Definition 2, Chapter IV, a square matrix A is unitary if, and only if,

$$F(Ax) = F(x),$$

in which

$$F(x) = \sum_{i=1}^{n} x_i \bar{x}_i, \ x = (x_i) \in C^n, \ 1 \le i \le n.$$

Consider the Hermitian form

$$H(x) = \sum_{i,j=1}^{n} a_{ij} x_i \bar{x}_j,$$

in which $a_{ij} = \bar{a}_{ji}$, $x = (x_i) \in C^n$, $1 \le i \le n$. $H(x)$ is said to be *positive-definite* if for each $x \in C^n$, $H(x) \ge 0$, and $H(x) = 0$ if, and only if, $x = 0$.

It is known (Stoll[43] or MacDuffee[31]) that a matrix A that leaves the positive-definite Hermitian form $H(x)$ invariant is similar to a unitary matrix. In other words, if $H(x)$ is the positive-definite Hermitian form and if A is a square matrix such that $H(Ax) = H(x)$ for each $x \in C^n$, then there exists a regular square matrix T such that $T^{-1}AT$ is unitary.

Definition 1.   (a) A matrix A is said to be *reducible* if

$$A = \begin{bmatrix} B & D \\ 0 & C \end{bmatrix},$$

in which B is a square matrix of order r $(0 < r < n)$, C a square matrix of order $n - r$, D and 0 are rectangular matrices, and 0 has all elements equal to zero.

(b) A matrix that is not reducible is called *irreducible*.

(c) Suppose

$$A = \begin{bmatrix} B_1 & & & & & 0 \\ & \cdot & & & & \\ & & \cdot & & & \\ & & & B_2 & \cdots & \\ & & & & \cdot & \\ 0 & & & & & 0 \\ & & & & & \cdot \\ 0 & \cdots & 0 & \cdots & B_r \end{bmatrix},$$

in which each B_i is a square irreducible matrix. Then A is said to be *completely reducible*.

Proposition 1.   *Let A be a square matrix of order n. A is reducible if, and only if, some linear subspace V of dimension r $(0 < r < n)$ in C^n remains invariant when A is regarded as a linear transformation of C^n.*

PROOF. Simple.

Proposition 2. *If A is a reducible square matrix so is its transpose A'.*

PROOF. Simple.

(d) A set $\mathcal{M}$ of square matrices of order $n \times n$ is said to be *reducible* if there exists a positive integer $r < n$ such that each matrix $M \in \mathcal{M}$, regarded as a linear transformation of C^n, maps C^r into itself. Otherwise, the set $\mathcal{M}$ is called *irreducible*. If the set $\mathcal{M}$ leaves C^r and its orthogonal complement C^{n-r} both invariant, then $\mathcal{M}$ is said to be *completely reducible*.

Theorem 1. (*Schur's Lemma.*) *Let $\mathcal{M}$ and $\mathcal{N}$ denote two irreducible sets of square matrices of order m and n, respectively. Let A be a matrix of order $m \times n$ such that for each $M \in \mathcal{M}$ there is some $N \in \mathcal{N}$, and for each $N_1 \in \mathcal{N}$ there is some $M_1 \in \mathcal{M}$ such that $MA = AN$ and $AM_1 = N_1 A$. Then either $A = 0$ or A is a regular square matrix.*

PROOF. Let $A = (a_{ij})$, $1 \leq i \leq m$, $1 \leq j \leq n$. If $a_k = (a_{1k}, \ldots, a_{mk})$, then $a_k \in C^m$ for each $k = 1, \ldots, n$. Let V be the subspace generated by a_k ($1 \leq k \leq n$). First we show that each M in $\mathcal{M}$ leaves V invariant.

Let $M = (u_{ij})$ and $N = (v_{ij})$. We compute MA:

$$Ma_k = \sum_{j=1}^{m} u_{ij} a_{jk} = b_{ik}, \quad \text{say.}$$

If $b_k = (b_{1k}, \ldots, b_{mk})$, then $b_k \in C^m$. On the other hand, since $MA = AN$,

$$b_{ik} = \sum_{j=1}^{n} a_{ij} v_{jk}.$$

Hence,

$$b_k = \left(\sum_{j=1}^{n} a_{1j} v_{jk}, \ldots, \sum_{j=1}^{n} a_{mj} v_{jk} \right)$$

$$= \sum_{j=1}^{n} v_{jk} a_j \in V.$$

This shows that M leaves V invariant. Since M is irreducible, either dim $V = 0$ or dim $V = m$. In the first case, $a_k = 0$ for each k. Hence, $A = 0$. In the second case, m out of n column vectors a_k ($1 \leq k \leq n$) are linearly independent. That means $m \leq n$.

Similarly, by taking transposes in $MA = AN$, one obtains $A'M' = N'A'$. Thus, arguing in the same way as above, one proves that either $A' = 0$ or $n \leq m$ and n out of m row vectors are linearly independent. Therefore, either $A = 0$ or A is a square matrix of order $n = m$. Since all column and row vectors of A are linearly independent, A is regular.

Corollary 1. *Let $\mathcal{M}$ be an irreducible set of square matrices of order n and let A be an $n \times n$ matrix such that $AM = MA$, for each $M \in \mathcal{M}$. Then $A = cI$, for some complex number c, in which I is the identity matrix.*

PROOF.   Let c be a complex number such that $|A - cI| = 0$, (i.e., c is a characteristic value of A). Let $B = A - cI$. Then

$$BM = (A - cI)M = AM - cM$$
$$MB = M(A - cI) = MA - cM.$$

By hypothesis $AM = MA$, so $BM = MB$. Since B is singular, $B = 0$ by the previous theorem. Hence, $A = cI$.

Corollary 2.   *Let $\mathcal{M}$ be an irreducible set of square matrices such that any two commute with each other. Then each M in $\mathcal{M}$ is a matrix of order 1.*

PROOF.   This is immediate from the above corollary, by identifying cI with c.

39.   ORTHOGONALITY RELATIONS

Definition 2.   (a) Let G be a topological group. A mapping f of G into $\mathbf{G}_n(C)$ (Notation 2, §27) is called a *representation* of G if f is a continuous homomorphism of G into $\mathbf{G}_n(C)$.
Clearly, $f(G)$ is a subgroup of $\mathbf{G}_n(C)$ and the kernel of f is an invariant closed subgroup of G. Hence $G/\ker f$ is isomorphic with $f(G)$.
(b) If $\ker f = \{e\}$, the representation is said to be *faithful*.
Observe that the smaller the kernel of f, the better the "approximation" of G by a full linear subgroup.

Notation. $\mathcal{M} = f(G)$, the set of all $n \times n$ matrices that are images of elements $x \in G$ under the representation f, will denote the representation f. Thus, for each $x \in G$, $f(x) = M_x \in \mathcal{M}$.

Since f is a homomorphism, $f(xy) = f(x)f(y)$ and $f(x^{-1}) = [f(x)]^{-1}$ for $x, y \in G$. This implies that $M_{xy} = M_x M_y$, $M_{x^{-1}} = M_x^{-1}$, where M_x, $M_y \in \mathcal{M}$.
(c) Let $\mathcal{M}$ and $\mathcal{N}$ be two representations of a topological group G. $\mathcal{M}$ and $\mathcal{N}$ are said to be *similar* if there exists a regular matrix A such that

$$N_x = A^{-1}M_x A, \text{ for each } x \in G, N_x \in \mathcal{N}, M_x \in \mathcal{M}.$$

The relation of being similar on the set of all representations of a topological group is called *similarity*.

Proposition 3.   *Similarity is an equivalence relation.*

PROOF.   Simple.
Remark.   The set of all representations of a topological group G can be decomposed into equivalence classes.

Definition 3. A representation $\mathcal{M}$ of a topological group is called *reducible* if $\mathcal{M}$ is reducible in the sense of Definition 1(d). A representation that is not reducible is called *irreducible*.

The next result shows that in each equivalence class of representations of a compact topological group there exists a special representation.

Theorem 2. *To each representation $\mathcal{M}$ of a compact Hausdorff topological group, there corresponds an equivalent unitary representation $\mathcal{U}$ (i.e., each U_x in $\mathcal{U}$ is a unitary matrix).*

PROOF. Let $u = \{u_1, u_2, \ldots, u_n\}$ be a column vector in C^n. Suppose $F(u) = \sum_{i=1}^{n} u_i \bar{u}_i$. For each $x \in G$, $M_x \in \mathcal{M}$. Let $M_x = (m_{ij}(x))$, $1 \leq i, j \leq n$. Then

$$F(M_x u) = \sum_{i=1}^{n} \left(\sum_{j=1}^{n} m_{ij}(x) u_j \right) \left(\sum_{k=1}^{n} \bar{m}_{ik}(x) \bar{u}_k \right)$$

$$= \sum_{i,j,k=1}^{n} m_{ij}(x) \bar{m}_{ik}(x) u_j \bar{u}_k.$$

Let $a_{jk}(x) = \sum_{i=1}^{n} m_{ij}(x) \bar{m}_{ik}(x)$. Then $a_{jk}(x) = \bar{a}_{kj}(x)$. Therefore, $F(M_x u)$ is a positive definite Hermitian form.

Now define

$$H(u) = \int_G F(M_x u)\, dx,$$

in which the integration is taken over the whole group G with respect to the Haar measure on G. It is clear that $H(u)$ is a positive definite Hermitian form. We show that $H(u)$ is invariant under each matrix $M_y \in \mathcal{M}$. For, using the invariance of the Haar integral, we have

$$H(M_y u) = \int_G F(M_x M_y u)\, dx$$

$$= \int_G F(M_{xy} u)\, dx$$

$$= \int_G F(M_x u)\, dx$$

$$= H(u).$$

We remarked above, every matrix M_x leaving $H(u)$ invariant can be transformed into a unitary matrix U_x. That is, there exists a matrix T (regular) such that

$$U_x = T^{-1} M_x T \qquad \text{for each } x \in G.$$

Clearly, $x \to U_x$ is a representation of G. This completes the proof.

Theorem 3. *Let $\mathcal{M}$ and $\mathcal{N}$ be two distinct irreducible inequivalent unitary representations of order m and n, respectively, of a compact Hausdorff topological group G. Then*

$$\int_G m_{ij}(x)\bar{n}_{kp}(x)\, dx = 0,$$

in which for each $x \in G$,

$$M_x = (m_{ij}(x)), \quad N_x = (n_{kp}(x)), \quad 1 \le i, \ j \le m, \ \ 1 \le k, \ p \le n.$$

PROOF. Let $B = (c_{ij})$ be an $m \times n$ matrix of constant coefficients. Consider

$$A_x = M_x B N_{x^{-1}}.$$

Let $A = \displaystyle\int_G A_x\, dx$. Then, using the invariance of the Haar measure, we obtain

$$M_y A N_{y^{-1}} = M_y \left(\int_G A_x\, dx\right) N_{y^{-1}}$$

$$= \int_G M_y A_x N_{y^{-1}}\, dx$$

$$= \int_G M_y M_x B N_{x^{-1}} N_{y^{-1}}\, dx$$

$$= \int_G M_{yx} B N_{(yx)^{-1}}\, dx$$

$$= \int_G A_x\, dx = A.$$

In other words, A commutes with all matrices in $\mathcal{M}$ and $\mathcal{N}$. Since these are irreducible representations, by Schur's lemma either $A = 0$ or A is a regular matrix of order $n \times n$. Since the representations are inequivalent, the latter case is ruled out. Hence,

$$A = 0 = \int_G M_x B N_{x^{-1}}\, dx$$

for any arbitrary matrix B of order $m \times n$. In particular, let B be the matrix in which all coefficients are zero except at jth row and pth column. Thus we may conclude that

$$\int_G m_{ij}(x)\bar{n}_{kp}(x)\, dx = 0$$

by observing that N_x is unitary.

This theorem says that the representation functions belonging to different inequivalent irreducible representations are orthogonal (Definition 4(b), §40).

Corollary 3. *If $\chi(x) = $ trace of M_x, $\chi'(x) = $ trace of N_x for each x in the above theorem, then*

$$\int_G \chi(x)\bar{\chi}'(x)\, dx = 0.$$

PROOF.   Since $\chi(x) = \sum_{i=1}^{n} m_{ii}(x)$, $\chi'(x) = \sum_{i=1}^{n} n_{ii}(x)$, the corollary follows from the theorem.

Theorem 4. *Let $\mathcal{M}$ be a unitary irreducible representation of a compact group G. Let $M_x = (m_{ij}(x)) \in \mathcal{M}$, $1 \le i, j \le n$. Then*

(i)
$$\int_G m_{ij}(x)\bar{m}_{ij}(x)\, dx = \frac{1}{n}$$

(ii)
$$\int_G m_{ij}(x)\bar{m}_{kp}(x)\, dx = 0 \qquad \text{for} \quad i \ne k, j \ne p.$$

PROOF.   Let $B = (c_{ij})$ be an $n \times n$ matrix with constant coefficients. Let $A_x = M_x B M_{x^{-1}}$. Then as in Theorem 3, $\int_G A_x\, dx = A$ is a constant matrix and $M_y A M_{y^{-1}} = A$. By Corollary 1, §38, it follows that $A = cI$, since $\mathcal{M}$ is irreducible. Thus, one obtains

(*)
$$\int_G M_x B M_{x^{-1}}\, dx = cI.$$

Equating traces on both sides, we obtain

$$\int_G \sum_{i,j,k=1}^{n} m_{ij}(x) c_{jk} \bar{m}_{ik}(x)\, dx = nc.$$

Since M is unitary, $\sum_{j,k=1}^{n} \delta_{jk} c_{jk} = nc$, in which $\delta_{ij} = 0$ or 1 according as $i \ne j$ or $i = j$.

If we choose B so that all entries are zero except the entry c_{jp}, which we set equal to 1, then $\delta_{jp} = nc$ or $c = \frac{1}{n}\delta_{jp}$. But then from (*) one obtains, by equating the coefficients at ith row and kth column,

$$\int_G m_{ij}(x)\bar{m}_{kp}(x)\, dx = \frac{1}{n}\delta_{jp}\,\delta_{ik}.$$

From this follow (i) and (ii), by using the definition of δ_{ij}.

In the notations of Corollary 3, one obtains the following:

Corollary 4. *For any irreducible unitary representation of a compact Hausdorff topological group G,*

(**)
$$\int_G \chi(x)\bar{\chi}(x)\, dx = 1.$$

PROOF. This follows from Theorem 4.

Remark. Let $\mathscr{M}$ be a representation of a topological group G. As above, let $\chi(x) = \sum_{i=1}^{n} m_{ii}(x)$, in which $M_x = (m_{ij}(x)) \in \mathscr{M}$. Then χ is called a *trace function* of G, and for $x, y \in G$,

$$\chi(x^{-1}yx) = \chi(y).$$

Let M be a reducible matrix such that

$$M = \begin{bmatrix} B_1 & \cdots & \cdots & \\ \cdot & & & \\ \cdot & B_2 & & 0 \\ \cdot & & & \\ & & & \cdot \\ 0 & & & \cdot \\ & & & \cdot \\ \cdot & & & \\ \cdot & \cdots & B_m \end{bmatrix},$$

where B_i's are irreducible matrices. Then

$$\text{Trace } M = \sum_{i=1}^{m} \text{Trace } B_i.$$

Thus, for any completely reducible representation $\mathscr{M}$ of G,

$$\chi(x) = \sum_{i=1}^{m} \chi_i(x),$$

in which $\chi_i(x)$ are the trace functions of the irreducible components of $\mathscr{M}$.

40. ORTHONORMAL FAMILY OF FUNCTIONS ON METRIZABLE COMPACT GROUPS

Definition 4. (a) Two complex-valued continuous functions f and g on a Hausdorff compact topological group G are said to be *orthogonal* if

$$\int_G f \bar{g} dx = 0.$$

(b) A family $\mathfrak{F}$ of complex-valued continuous functions on G is said to be an *orthogonal family* if each pair $f, g (f \neq g)$ of functions in $\mathfrak{F}$ is orthogonal.

(c) An orthogonal family $\mathfrak{F}$ is said to be *orthonormal* if for each $f \in \mathfrak{F}$,

$$\int_G |f|^2 dx = 1.$$

Remark. Every orthogonal family (not containing the zero function) can be transformed into an orthonormal one. For example, for each $f \in \mathfrak{F}$, let $f' = \alpha f$, where $\alpha^{-2} = \int_G |f|^2 \, dx$. Then $\mathfrak{F}' = \{f'\}$ is an orthonormal family.

(d) Let E be a real or complex vector space with a topology, and F a subset of E. Then F is said to be *total* in E if the vector space generated by F is dense in E. Or, in other words, the set of all finite linear combinations of elements in F with real or complex coefficients is dense in E.

Theorem 5. *Let G be a metrizable compact topological group. Let $C(G)$ be the vector space of all continuous complex-valued functions on G and let $C(G)$ be endowed with the sup norm topology. Let $\{\varphi_m\}$ be a total, countable orthonormal family of complex-valued continuous functions in $C(G)$. Then for each $f \in C(G)$,*

$$\int_G |f|^2 \, dx = \sum_{m=1}^{\infty} |c_m|^2,$$

where $c_m = \int_G f \bar{\varphi}_m \, dx$ for each $m \geq 1$.

PROOF. Let $f_n = \sum_{m=1}^{n} c_m \varphi_m$. Then, clearly,

$$\int_G f_n \bar{f} \, dx = \sum_{m=1}^{n} c_m \int_G \bar{f} \varphi_m \, dx = \sum_{m=1}^{n} |c_m|^2$$

and

$$\int_G \bar{f}_n f \, dx = \sum_{m=1}^{n} |c_m|^2.$$

Furthermore,

$$0 \leq \int_G |f - f_n|^2 \, dx = \int_G (f - f_n)(\bar{f} - \bar{f}_n) \, dx$$

$$\leq \int_G |f|^2 \, dx - \int_G (f_n \bar{f} + f \bar{f}_n) \, dx + \int_G |f_n|^2 \, dx$$

$$\leq \int_G |f|^2 \, dx - \sum_{m=1}^{n} |c_m|^2.$$

Therefore, $\sum_{m=1}^{n} |c_m|^2 \leq \int_G |f|^2 \, dx$ for every n, and

(5.1) $$\sum_{m=1}^{\infty} |c_m|^2 \leq \int_G |f|^2 \, dx.$$

Now to show the reverse inequality, we make use of the fact that $\{\varphi_n\}$ is total.

Let $a_1, \ldots, a_n$ be a finite subset of complex numbers and put $g_n = \sum_{m=1}^{n} a_m \varphi_m$. Let $\alpha = \int_G |f - g_n|^2 \, dx \geq 0$. As above,

$$\alpha = \int_G |f|^2 \, dx - \int_G (f\bar{g}_n + \bar{f}g_n) \, dx + \int_G |g_n|^2 \, dx.$$

But

$$c_m = \int_G f\bar{\varphi}_m \, dx,$$

so

$$\int_G f\bar{g}_n \, dx = \sum_{m=1}^{n} \bar{a}_m c_m,$$

$$\int_G \bar{f}g_n \, dx = \sum_{m=1}^{n} a_m \bar{c}_m,$$

and also

$$\int_G |g_n|^2 \, dx = \sum_{m=1}^{n} |a_m|^2.$$

Whence,

$$\alpha = \int_G |f|^2 \, dx - \sum_{m=1}^{n} (c_m \bar{a}_m + \bar{c}_m a_m) + \sum_{m=1}^{n} |a_m|^2$$

$$= \int_G |f|^2 \, dx - \sum_{m=1}^{n} |c_m|^2 + \sum_{m=1}^{n} |c_m - a_m|^2.$$

This shows that α is minimum if $a_m = c_m$ for $m = 1, \ldots, n$, and

$$\min \alpha = \int_G |f|^2 \, dx - \sum_{m=1}^{n} |c_m|^2.$$

Since $\{\varphi_n\}$ is total in $C(G)$, for each $\varepsilon > 0$ there exists a set $\{a_1, \ldots, a_n\}$ of complex numbers such that $|f(x) - \sum_{m=1}^{n} a_m \varphi_m(x)| < \varepsilon$ for all $x \in G$. Hence,

$$\min \alpha \leq \int_G \left| f - \sum_{m=1}^{n} a_m \varphi_m \right|^2 dx \leq \varepsilon^2 \int_G dx = \varepsilon^2,$$

and, therefore,

$$\int_G |f|^2 \, dx - \sum_{m=1}^{n} |c_m|^2 \leq \varepsilon^2,$$

or

$$\int_G |f|^2 \, dx \leq \varepsilon^2 + \sum_{m=1}^{\infty} |c_m|^2.$$

But ε being arbitrary, we have

(5.2) $$\int_G |f|^2 \, dx \leq \sum_{m=1}^{\infty} |c_m|^2.$$

By combining (5.1) and (5.2), we have established the theorem.

Theorem 6. *Let* $\mathfrak{F} = \{\varphi_\alpha\}$ *be an orthogonal system of continuous complex-valued functions defined on a metrizable compact semitopological group G. Then* $\mathfrak{F}$ *is at most countable.*

PROOF. Without loss of generality we can assume that $\mathfrak{F}$ is an orthonormal family.

In view of Corollary 6, §17, Chapter II, G is a metrizable topological group. Since every metrizable compact topological space is separable, so is G and, hence, it satisfies the second axiom of countability.

Let $\{B_n\}(n \geq 1)$ be a countable base of the topology of G. Let $\{(U_n, V_n)\}$ denote a countable subfamily of pairs taken from $\{B_n\}$ such that $\bar{V}_n \subset U_n$ for each n. This is possible in view of the fact that G is normal. Since $\bar{V}_n$ is compact, there exists a continuous real-valued function g_n such that $0 \leq g_n \leq 1$, $g_n(\bar{V}_n) = 1$ and $g_n(G \sim U_n) = 0$ (Corollary 1, §26, Chapter IV).

Let φ_α be any function in $\mathfrak{F}$. Then $\varphi_\alpha = \varphi_{1\alpha} + i\varphi_{2\alpha}$, in which $i = \sqrt{-1}$ and $\varphi_{1\alpha}$, $\varphi_{2\alpha}$ are real-valued. Since φ_α is continuous and $\neq 0$, $\varphi_{1\alpha}$ and $\varphi_{2\alpha}$ are continuous and at least one of them is $\neq 0$. Let W be an open set on which $\varphi_{1\alpha}$ or $\varphi_{2\alpha}$ does not change sign, i.e., either $W = \{x \in G : \varphi_{1\alpha}(x) \text{ or } \varphi_{2\alpha}(x) > 0\}$ or $W = \{x \in G : \varphi_{1\alpha}(x) \text{ or } \varphi_{2\alpha}(x) < 0\}$. There exists an integer $m > 0$ such that $U_m \subset W$. Then

$$\int_G g_m \bar{\varphi}_\alpha \, dx = \int_{U_m} g_m \bar{\varphi}_\alpha \, dx \neq 0$$

because $g_m(U_m) > 0$, $\varphi_{1\alpha}$ or $\varphi_{2\alpha} \neq 0$ on U_m, and $\mu(U_m) > 0$.

For each n and k (positive integers), let

$$\mathfrak{F}_{n,k} = \left\{ \varphi_\alpha \in \mathfrak{F} : \left| \int_G g_n \bar{\varphi}_\alpha \, dx \right| > \frac{1}{k} \right\}.$$

If $c_\alpha = \int g_n \bar{\varphi}_\alpha \, dx$, then from the first part of Theorem 5 it follows that for any finite family of distinct α's,

$$\sum |c_\alpha|^2 \leq \int_G |g_n|^2 \, dx.$$

But since $\frac{1}{k} < \left| \int_G g_n \bar{\varphi}_\alpha \, dx \right|$ for each $\varphi_\alpha \in \mathfrak{F}_{n,k}$,

$$\sideset{}{'}\sum_\alpha k^{-2} \leq k^{-2} \sideset{}{'}\sum 1 \leq \int_G |g_n|^2 \, dx,$$

where $\sum'$ is taken over all α's such that $\varphi_\alpha \in \mathfrak{F}_{n,k}$. Thus,

$$\sideset{}{'}\sum_\alpha 1 \leq k^2 \int_G |g_n|^2 \, dx.$$

Since the right-hand side of the above inequality is a finite real number, $\sum'_\alpha 1$ is finite and, hence, $\mathfrak{F}_{n,k}$ consists of a finite family of functions for each pair of n and k. We show that $\mathfrak{F} = \bigcup_{n,k=1} \mathfrak{F}_{n,k}$. For any $\varphi_\alpha \in \mathfrak{F}$, there exists a positive integer m such that $\int g_m \bar{\varphi}_\alpha \, dx \neq 0$, as shown above. Therefore, there exists a sufficiently large k for which $\left| \int_G g_m \bar{\varphi}_\alpha \, dx \right| > \dfrac{1}{k}$, i.e., $\varphi_\alpha \in \mathfrak{F}_{n,k}$. This completes the proof.

Remark. Theorems 5 and 6 show that, for a metrizable compact group G, $C(G)$ is a separable pre-Hilbert space with $\int_G f\bar{g} \, dx$ as its scalar product.

41. INTEGRAL EQUATIONS ON COMPACT GROUPS

In this section, we shall prove two results, Theorems 7 and 8, concerning the integral equations used in proving Peter-Weyl's theorem (see next section).

We shall assume that G is a compact metrizable semitopological group. Let $C_R(G)$ denote the set of all continuous real-valued functions on G. Then $C_R(G)$ is a real pre-Hilbert space whose completion is $L_2(G)$. Let $k(x, y)$ denote a continuous real-valued function on the product $G \times G$ and assume $k(x, y) = k(y, x)$, i.e., k is a symmetric continuous function on $G \times G$. Let $g \in C_R(G)$. We seek $f \in C_R(G)$ such that either

$$(1) \qquad g(x) = \int_G k(x, y)f(y) \, dy \qquad \text{for all } x \in G,$$

or

$$(2) \qquad f(x) - g(x) = \int_G k(x, y)f(y) \, dy \qquad \text{for all } x \in G.$$

If $g(x) = 0$ in (2), then

$$(3) \qquad f(x) = \int_G k(x, y)f(y) \, dy.$$

In the classical case when G is replaced by a closed bounded interval $I = [a, b]$, equations (1) and (2) are called the *Fredholm integral equations* of the *first* and *second kind*, respectively. Equation (3) is called the homogeneous Fredholm integral equation of the second kind. (All integrals are with respect to the Haar measure on G. After having normalized the measure, we shall assume that $\mu(G) = 1$.)

Consider the following:

$$(4) \qquad f(x) = \lambda \int_G k(x, y)f(y) \, dy,$$

in which $k(x, y)$ is real-valued, continuous and $k(x, y) = k(y, x)$. This equation is called an integral equation with a parameter λ. A function f satisfying (4) for a given λ is called a *characteristic* function and λ a *characteristic value* of the *kernel* $k(x, y)$.

It is known (Exercise 7(e) at the end of this chapter) that each symmetric nonidentically zero kernel has a nonzero real or complex characteristic value.

If $k(x, y)$ is real, then all characteristic values are real and characteristic functions belonging to different λ's are real-valued and orthogonal on I. If $\lambda = 0$, it is clear that the only solution of (4) is $f \equiv 0$. Thus, the set of all solutions of (4), when $\lambda = 0$, is a zero-dimensional vector space. In the sequel we shall assume that $\lambda \neq 0$.

Let $\lambda \neq 0$ be a real number. Let V_λ denote the set of all f which satisfy (4). Owing to the linearity of the integral, it is quite clear that V_λ forms a real vector space for a given λ. First of all, we show the following:

Theorem 7. *For a given λ, V_λ is finite-dimensional.*

Proof. Let $\{\varphi_i\}$ $(1 \leq i \leq n)$ be a finite orthonormal family in V_λ and consider

$$k^*(x, y) = \lambda^{-1} \sum_{i=1}^{n} \varphi_i(x)\varphi_i(y), \qquad (x, y \in G).$$

Since the φ_i are orthonormal $\left(\text{i.e., } \int_G \varphi_i\varphi_j = \delta_{ij}\right)$, it is easy to see that

$$\int_G \int_G k^{*2}(x, y) \, dx \, dy = n\lambda^{-2}.$$

Now, since $[k(x, y) - k^*(x, y)]^2 \geq 0$, we obtain

$$0 \leq \int_G \int_G [k(x, y) - k^*(x, y)]^2 \, dx \, dy$$

$$\leq \int_G \int_G k^2(x, y) \, dx \, dy - n\lambda^{-2}.$$

That means

$$n \leq \lambda^2 \int_G \int_G k^2(x, y) \, dx \, dy.$$

Since k^2 is continuous and G is compact, the right-hand integral is finite. This shows that the number of orthonormal functions in V_λ cannot exceed a finite number. Hence, by the Gram-Schmidt construction (Stoll[43]) V_λ has a finite orthonormal basis. In other words, V_λ is finite-dimensional.

Remark. Theorem 7 follows from a more general result, viz., Exercise 9(iii).

Observe that if $\lambda_1 \neq \lambda_2$, then the characteristic functions φ_{λ_1} and φ_{λ_2} of a symmetric continuous kernel $k(x, y)$, corresponding to characteristic values λ_1 and λ_2, respectively, are orthogonal i.e., $\int_G \varphi_{\lambda_1}(x)\varphi_{\lambda_2}(x) \, dx = 0$. Thus, it follows that the corresponding vector spaces V_{λ_1} and V_{λ_2} are mutually orthogonal. Since each vector space V_λ is finite-dimensional by Theorem 7, V_λ can contain only a finite orthonormal family of characteristic functions. Enumerating the entire set of orthonormal characteristic functions, we obtain a countable orthonormal family $\{\varphi_i\}$ of characteristic functions. We see that the orthonormal family $\{\varphi_i\}$ satisfies the following properties:

(a) Each φ_i is a continuous real-valued function on G.

(b) $\displaystyle\int_G \varphi_i(x)\varphi_j(x)\, dx = \delta_{ij}$.

(c) Each φ_i is a solution of equation (4) for some real λ_i.

Before we prove our next result, we have a few lemmas.

Lemma 1. *Let $\{\varphi_i\}$ be orthonormal characteristic functions with corresponding characteristic values λ_i. Then for any positive integer n,*

$$\sum_{i=1}^{n} \lambda_i^{-2}\varphi_i^2 \le M^2$$

uniformly on G, where $M > 0$ and is independent of n and x.

PROOF. In view of the inequality (5.1) in the proof of Theorem 5, §40, by putting $f(x) = k(x, y)$ for any fixed y, we have

$$\sum_{i=1}^{n} \left[\int_G k(x, y)\varphi_i(y)\, dy \right]^2 \le \int_G k^2(x, y)\, dy$$

for each positive integer n and $x \in G$. Since $k(x, y)$ is continuous on the compact space $G \times G$, we have $|k(x, y)| \le M$ which is independent of $x, y \in G$. Thus, by using equation (4), we have

$$\sum_{i=1}^{n} \lambda_i^{-2}\varphi_i^2 \le M^2 \qquad (\text{since } \mu(G) = 1).$$

Lemma 2.

$$\sum_{i=1}^{\infty} \lambda_i^{-2} < \infty.$$

PROOF. By Lemma 1,

$$\sum_{i=1}^{n} \lambda_i^{-2}\varphi_i^2 \le M^2.$$

But then by integrating we have

$$\sum_{i=1}^{n} \lambda_i^{-2}\int_G \varphi_i^2(x)\, dx \le M^2.$$

Using orthonormality of φ_i's, we have

$$\sum_{i=1}^{n} \lambda_i^{-2} \le M^2$$

for each n. By letting $n \to \infty$, we see that the lemma is established.

Lemma 3. *The series*

(a)
$$\sum_{i=1}^{\infty} \lambda_i^{-4} \varphi_i(x) \varphi_i(y)$$

and

(b)
$$\sum_{i=1}^{\infty} \lambda_i^{-2} \varphi_i(x) \varphi_i(y)$$

are uniformly convergent on $G \times G$ and, hence, represent continuous functions.

PROOF. (a) Consider

$$|S_{nm}(x, y)| = \left| \sum_{n \leq i \leq m} \lambda_i^{-4} \varphi_i(x) \varphi_i(y) \right|$$

$$= \left| \sum_{n \leq i \leq m} \lambda_i^{-2} [\lambda_i^{-1} \varphi_i(x)][\lambda_i^{-1} \varphi_i(y)] \right|$$

$$\leq \left[\sum_{n \leq i \leq m} |\lambda_i^{-1} \varphi_i(x)| \, |\lambda_i^{-1} \varphi_i(y)| \right] \left[\sum_{n \leq i \leq m} \lambda_i^{-2} \right].$$

Using Schwarz's inequality (Titchmarsh[44]), for the first summation we have

$$|S_{nm}(x, y)| \leq \sqrt{\sum_{n \leq i \leq m} \lambda_i^{-2} \varphi_i^2(x)} \sqrt{\sum_{n \leq i \leq m} \lambda_i^{-2} \varphi_i^2(y)} \left(\sum_{n \leq i \leq m} \lambda_i^{-2} \right).$$

By Lemma 1, we have

$$|S_{nm}(x, y)| \leq M^2 \sum_{n \leq i \leq m} \lambda_i^{-2}.$$

Since $\sum_{i=1}^{\infty} \lambda_i^{-2}$ is convergent by Lemma 2, we see by the general principle of uniform convergence that (a) is uniformly convergent. The uniform convergence of (b) follows similarly.

Lemma 4. *Let $k(x, y)$ be a symmetric continuous real-valued function on $G \times G$. For an integer $n > 1$, define*

$$k_n(x, y) = \int_G k_{n-1}(x, t) k(t, y) \, dt$$

by iteration, in which $k_1(x, y) = k(x, y)$. Then $k_n(x, y)$ is a continuous symmetric function on $G \times G$. If $\{\lambda_i\}$ is the set of characteristic values of k then $\{\lambda_i^n\}$ is the set of characteristic values of k_n.

PROOF. It is clear that $k_n(x, y)$ is continuous. The fact that k_n is symmetric follows by induction, since k is symmetric. Let $\{\varphi_i\}$ denote the characteristic functions corresponding to λ_i. Then

$$\varphi_i(x) = \lambda_i \int_G k(x, y) \varphi_i(y) \, dy.$$

Multiplying the above equation by $k(t, x)$ and then integrating with respect to x, we obtain

$$\varphi_i(t) = \lambda_i \int_G k(t, x)\varphi_i(x)\, dx = \lambda_i^2 \int_G \int_G k(t, x)k(x, y)\varphi_i(y)\, dx\, dy$$

$$= \lambda_i^2 \int_G k_2(t, y)\varphi_i(y)\, dy.$$

This shows that λ_i^2 is a characteristic value of $k_2(x, y)$. By induction on n, we see that λ_i^n is a characteristic value of $k_n(x, y)$. We ask the reader to show that $k_n(x, y)$ has no characteristic values other than λ_i^n, and each characteristic function belonging to $k_n(x, y)$ is a linear combination of $\{\varphi_i\}$ (Lovitt,[30] p. 146).

Before we prove Lemma 5, we observe that if k is a symmetric continuous nonzero real-valued kernel then it has at least one nonzero characteristic value (Exercise 7(e) at the end of this chapter). This means that if a symmetric continuous kernel does not have a nonzero characteristic value then it must be identically equal to zero.

Lemma 5. *If the series*

$$\sum_{i=1}^{\infty} \lambda_i^{-1}\varphi_i(x)\varphi_i(y)$$

is uniformly convergent, then

$$k(x, y) = \sum_{i=1}^{\infty} \lambda_i^{-1}\varphi_i(x)\varphi_i(y).$$

PROOF. Put $\tilde{k}(x, y) = \sum_{i=1}^{\infty} \lambda_i^{-1}\varphi_i(x)\varphi_i(y).$

Then $\tilde{k}$ is a real-valued (since each φ_i is real-valued) continuous (since the series is a uniformly convergent series of continuous functions) symmetric function on $G \times G$. We show that $h(x, y) \equiv k(x, y) - \tilde{k}(x, y)$ is identically zero. If $h \neq 0$, there exists at least one characteristic value $\lambda \neq 0$. Let φ be a nonzero characteristic function of $h(x, y)$. Then

$$\varphi(x) = \lambda \int_G h(x, y)\varphi(y)\, dy.$$

Substituting $k(x, y) - \tilde{k}(x, y)$ for $h(x, y)$, we obtain

(1) $$\varphi(x) = \lambda \int_G k(x, y)\varphi(y)\, dy - \lambda \int_G \sum_{i=1}^{\infty} \lambda_i^{-1}\varphi_i(x)\varphi_i(y)\varphi(y)\, dy.$$

Multiplying the last equation by $\varphi_j(x)$ and integrating with respect to x, we have

$$\int_G \varphi(x)\varphi_j(x)\, dx = \lambda \int_G \int_G k(x, y)\varphi(y)\varphi_j(x)\, dx\, dy$$

$$- \lambda \int_G \int_G \sum_{i=1}^{\infty} \lambda_i^{-1}\varphi_i(y)\varphi_i(x)\varphi_j(x)\varphi(y)\, dx\, dy.$$

By orthonormality of $\{\varphi_i\}$, the last equation reduces to

$$(2) \qquad \int_G \varphi(x)\varphi_j(x)\,dx = \lambda \int_G \varphi(y)\,dy \int_G k(x,y)\varphi_j(x)\,dx - \frac{\lambda}{\lambda_j} \int \varphi_j(y)\varphi(y)\,dy$$

$$= \frac{\lambda}{\lambda_j} \int_G \varphi_j(y)\varphi(y)\,dy - \frac{\lambda}{\lambda_j} \int \varphi_j(y)\varphi(y)\,dy$$

$$= 0 \quad \text{for all } j \geq 1.$$

Hence from (1) we obtain,

$$\varphi(x) = \lambda \int k(x,y)\varphi(y)\,dy.$$

That means that φ is a characteristic function of $k(x,y)$. Hence φ can be written as a linear combination of orthonormal functions $\{\varphi_i\}$. Let

$$\varphi(x) = \sum_{i=1}^{n} \alpha_i \varphi_i(x)$$

for some real numbers α_i. By (2) we know that

$$0 = \int_G \varphi(x)\varphi_j(x)\,dx = \sum_{i=1}^{n} \alpha_i \int \varphi_i(x)\varphi_j(x)\,dx$$

$$= \alpha_j \quad \text{for all } j \geq 1.$$

Hence, φ is identically zero. This is a contradiction, since $\varphi \neq 0$. Hence, $h = 0$.

Lemma 6. *If for a $p \in C_R(G)$ and all $n \geq 1$,*

$$\int_G p\varphi_n \, dx = 0,$$

then

$$\int_G k(x,y)p(y)\,dy = 0 \qquad \text{for all } x \in G.$$

PROOF. Since the series (a) in Lemma 3 is uniformly convergent, $k_4(x,y) = \sum \lambda_i^{-4}\varphi_i(x)\varphi_i(y)$ by Lemmas 4 and 5. Multiplying both sides by $p(x)p(y)$ and integrating termwise, we have

$$(1) \qquad \int_G \int_G k_4(x,y)p(x)p(y)\,dx\,dy$$

$$= \sum_{i=1}^{\infty} \lambda_i^{-4} \int_G \varphi_i(x)p(x)\,dx \int_G \varphi_i(y)p(y)\,dy = 0$$

$$\text{(by assumption).}$$

Furthermore, $k_2(x,y) = \sum_{i=1}^{\infty} \lambda_i^{-2}\varphi_i(x)\varphi_i(y)$, by Lemmas 3, 4 and 5; and by orthonormality of $\{\varphi_i\}$, we have

$$\text{(2)} \qquad \int_G k_2(x, z)k_2(y, z) \, dz = k_4(x, y).$$

From (1), by substituting (2), we have

$$\int_G \int_G \int_G k_2(x, z)k_2(y, z)p(x)p(y) \, dx \, dy \, dz = 0,$$

which is the same as

$$\int_G dz \left[\int_G k_2(x, z)p(x) \, dx \int_G k_2(y, z)p(y) \, dy \right] = 0,$$

or

$$\int_G dz \left[\int_G k_2(x, z)p(x) \, dx \right]^2 = 0.$$

Since $\int_G k_2(x, z)p(x) \, dx$ is a continuous function of z, we have

$$\int_G k_2(x, z)p(x) \, dx = 0$$

for all $z \in G$. Thus, by putting $z = y$, we have

$$\text{(3)} \qquad \int_G k_2(x, y)p(x) \, dx = 0$$

for all $y \in G$.

Multiplying (3) by $p(y)$ and then integrating it, we obtain

$$\int_G \int_G k_2(x, y)p(x)p(y) \, dx \, dy = 0.$$

Since

$$k_2(x, y) = \int_G k(x, z)k(y, z) \, dz \qquad \text{(Lemma 4),}$$

repeating the same argument as for $k_4(x, y)$, we obtain

$$\int_G k(x, y)p(x) \, dx = 0$$

for all $y \in G$. In view of the symmetry of k, we have thus established the lemma.

Now we have the following:

Theorem 8. *Let $\{\varphi_i\}$ denote the countable family of orthonormal characteristic functions of (4) and let λ_i be the characteristic value corresponding to φ_i. If for some $f \in C_R(G)$,*

$$\text{(8.1)} \qquad g(x) = \int_G k(x, y)f(y) \, dy,$$

then

(8.2)
$$g = \sum_{i=1}^{\infty} \alpha_i \varphi_i,$$

in which $\alpha_i = \int_G g\varphi_i \, dx$. *The series* (8.2) *converges uniformly on G.*

PROOF.   (a) We first show that the series (8.2) is uniformly convergent. For this, consider

$$S_{nm} = \sum_{n \leq i \leq m} \alpha_i \varphi_i.$$

By definition and using (8.1), we have

$$\alpha_i = \int_G g(x)\varphi_i(x) \, dx$$

$$= \int_G \left[\int_G k(x, y)f(y) \, dy \right] \varphi_i(x) \, dx$$

$$= \int_G f(y) \left[\int_G k(x, y)\varphi_i(x) \, dx \right] dy \qquad \text{(by Theorem 5, §34)}$$

$$= \int_G f(y)\lambda_i^{-1}\varphi_i(y) \, dy = \lambda_i^{-1} \int_G f(y)\varphi_i(y) \, dy$$

$$= \lambda_i^{-1}\beta_i,$$

in which

$$\beta_i = \int_G f\varphi_i \, dx.$$

Therefore,

$$|S_{nm}| \leq \sum_{n \leq i \leq m} |\alpha_i| \, |\varphi_i|$$

$$\leq \sum_{n \leq i \leq m} |\lambda_i^{-1}\beta_i| \, |\varphi_i|$$

$$\leq \sum_{n \leq i \leq m} |\beta_i| \, |\lambda_i^{-1}\varphi_i|$$

$$\leq \sqrt{\sum_{n \leq i \leq m} |\beta_i|^2} \sqrt{\sum_{n \leq i \leq m} \lambda_i^{-2}\varphi_i^2}$$

by Schwarz's inequality (Titchmarsh[44]).

But then by Lemma 1,

$$|S_{nm}| \leq M \sqrt{\sum_{n \leq i \leq m} |\beta_i|^2}.$$

Furthermore, by Theorem 5, §40, we have

$$\sum_{i=1}^{\infty} |\beta_i|^2 \leq \int_G |f|^2 \, dx,$$

which is a convergent series.   Hence S_{nm} converges to 0 (as $n, m \to \infty$) uniformly in x.   Therefore, (8.2) is a uniformly convergent series.

(b) Now we show that

$$g = \sum_{i=1}^{\infty} \alpha_i \varphi_i.$$

By part (a), $h = \sum_{j=1}^{\infty} \alpha_i \varphi_i$ is a continuous real-valued (because each φ_i is real-valued and each α_i is real) function, because the series is uniformly convergent. In order to establish (b), we have to show that $h \equiv g$. For this, it is sufficient to show that the real-valued continuous function $p = g - h$ is identically zero. Since p is continuous and real-valued, to show that $p = 0$, it is sufficient to show that

$$\int_G p^2 \, dx = 0.$$

From $p = g - h$ one obtains

$$p^2 = gp - hp$$

and, hence,

(8.3)
$$\int_G p^2 \, dx = \int_G gp \, dx - \int_G hp \, dx$$
$$= \int_G gp \, dx - \sum_{i=1}^{\infty} \alpha_i \int_G p \varphi_i \, dx,$$

since (8.2) is uniformly convergent. But from $p = g - \sum_{i=1}^{\infty} \alpha_i \varphi_i$, one obtains

(8.4)
$$\int_G p \varphi_j \, dx = \int_G g \varphi_j \, dx - \sum_{i=1}^{\infty} \alpha_i \int_G \varphi_i \varphi_j \, dx$$
$$= \alpha_j - \sum_{i=1}^{\infty} \alpha_i \delta_{ij} = \alpha_j - \alpha_j = 0$$

for all $j \geq 1$. Hence by Lemma 6,

(8.5)
$$\int_G k(x, y) p(y) \, dy = 0 \qquad \text{for all } x \in G.$$

By virtue of (8.4) and (8.5), (8.3) yields

$$\int_G p^2 \, dx = \int_G gp \, dx = \int_G \int_G k(x, y) f(y) \, dy \, p(x) \, dx$$
$$= \int_G f(y) \left[\int_G k(x, y) p(x) \, dx \right] dy = 0.$$

This completes the proof.

Remark. Theorem 8 follows from a more general result, viz., Exercise 7(f).

42. THE PETER-WEYL THEOREM

We prove the following two lemmas from which we shall derive the theorem of Peter-Weyl. The proofs of the lemmas are based on Pontrjagin.[37]

Lemma 7. *Let G be a compact metrizable semitopological group. Let C(G) denote the set of all continuous complex-valued functions on G, and let C(G) be endowed with the sup norm topology. Let $\mathfrak{F}_0$ denote the set of all continuous complex-valued functions φ on G such that for some symmetric, real-valued continuous function k $(k(x) = k(x^{-1}))$ on G and $\lambda \neq 0$,*

(i) $$\varphi(x) = \lambda \int_G k(x^{-1}y)\varphi(y)\,dy.$$

Then $\mathfrak{F}_0$ is total in C(G).

PROOF. Observe that G is a compact topological group. Let $f \in C(G)$. G being compact, f is uniformly continuous. For $\varepsilon > 0$, there exists a symmetric neighborhood U of $e \in G$ such that for all $x, y \in G$, $x^{-1}y \in U$ implies

(ii) $$|f(x) - f(y)| < \varepsilon/2.$$

Let V be a neighborhood of e such that $\bar{V} \subset U$. There exists a real continuous function g on G such that $0 \leq g \leq 1$, $g(\bar{V}) = 1$ and $g(G \sim U) = 0$.

Let $k(x) = \alpha(g(x) + g(x^{-1}))$, where α is chosen in such a way that $\int_G k(x)\,dx = 1$. Clearly, k is a symmetric continuous function on G. Set

$$h(x) = \int_G k(x^{-1}y)f(y)\,dy.$$

Since

$$f(x) = f(x)\int_G k(x^{-1}y)\,dy = \int_G k(x^{-1}y)f(x)\,dy,$$

owing to the choice of $k(x)$ and the invariance of the Haar integral, we have

(iii) $$|h(x) - f(x)| < \varepsilon/2,$$

by using (ii). Hence by Theorem 8, §41, $h(x) = \sum_{n=1}^{\infty} \alpha_n \varphi_n(x)$, where the series is uniformly convergent and where for each $n \geq 1$,

(iv) $$\varphi_n(x) = \lambda_n \int_G k(x^{-1}y)\varphi_n(y)\,dy$$

for some number $\lambda_n \neq 0$. Therefore, for sufficiently large m, we have

(v) $$\left| h(x) - \sum_{n=1}^{m} \alpha_n \varphi_n(x) \right| < \varepsilon/2$$

for all $x \in G$. Now combining (iii) and (v), we obtain

$$\left| f(x) - \sum_{n=1}^{m} \alpha_n \varphi_n(x) \right| < \varepsilon$$

for all $x \in G$. Since f and φ_n are all continuous, it follows that

$$\left\| f - \sum_{n=1}^{m} \alpha_n \varphi_n \right\|_{\infty} = \sup_{x \in G} \left| f(x) - \sum_{n=1}^{m} \alpha_n \varphi_n(x) \right| \leq \varepsilon.$$

Now since the φ_n are solutions of (iv), which is of the form (i), the proof of the lemma is completed.

Lemma 8. *Let G be a metrizable compact semitopological group. Let $\mathfrak{F}$ denote the totality of all functions belonging to representations of G. (In other words, if $\mathcal{M} = \{M_x\}$ is a representation of G, where $M_x = (m_{ij}(x))$ is a regular matrix, then $m_{ij} \in \mathfrak{F}$.) Then each function in $\mathfrak{F}_0$ (Lemma 7) is a linear combination of functions in $\mathfrak{F}$.*

PROOF. Let $\varphi \in \mathfrak{F}_0$. Then there exists a $\lambda \neq 0$ and a continuous kernel k such that

(i) $$\varphi(x) = \lambda \int_G k(x^{-1}y)\varphi(y)\,dy.$$

The vector space V_λ of all such φ's for a given λ, is finite-dimensional by Theorem 7, §41. Let $\{\varphi_i\}$ $(1 \leq i \leq n)$ denote a basis of orthonormal solutions of (i). Using the invariance of the Haar integral, for any $z \in G$, we have

$$\varphi_i(zx) = \lambda \int_G k(x^{-1}z^{-1}y)\varphi_i(y)\,dy$$

$$= \lambda \int_G k(x^{-1}z^{-1}zy)\varphi_i(zy)\,dy$$

$$= \lambda \int_G k(x^{-1}y)\varphi_i(zy)\,dy.$$

This shows that $\varphi_i(zx)$ $(1 \leq i \leq n)$ is also a solution of (i) and, therefore,

(ii) $$\varphi_i(zx) = \sum_{j=1}^{n} m_{ij}(z)\varphi_j(x),$$

in which $m_{ij}(z)$ are complex numbers depending on z. Let $M_z = (m_{ij}(z))$ denote the $n \times n$ matrix. Owing to invariance of the Haar integral and the orthonormality of the $\{\varphi_i\}$,

$$\int_G \varphi_i(zx)\varphi_j(zx)\,dx = \int_G \varphi_i(x)\varphi_j(x)\,dx = \delta_{ij}.$$

Hence, the translates $\{\varphi_i(zx)\}$ of $\{\varphi_i(x)\}$ also form a linearly independent system of orthonormal functions. Therefore, $\{\varphi_i(zx)\}$ $(1 \leq i \leq n)$ form a basis of V_λ and so M_z is a regular matrix. Since z is arbitrary, $\mathcal{M} = \{M_z\}$, $z \in G$, is a set of regular $n \times n$ matrices. We show that the mapping: $z \to M_z$ is a continuous homomorphism.

Multiplying (ii) by $\varphi_k(x)$ and integrating, we have

$$\int_G \varphi_i(zx)\varphi_k(x)\,dx = \sum_{j=1}^{n} m_{ij}(z)\int_G \varphi_j(x)\varphi_k(x)\,dx$$

$$= \sum_{j=1}^{n} m_{ij}(z)\,\delta_{jk} = m_{ik}(z).$$

Since the φ_i are continuous functions, $m_{ik}(z)$ is also continuous and, therefore, the mapping: $z \to M_z$ is continuous.

Furthermore, from (ii) again it follows that

$$\varphi_i(zyx) = \sum_{j=1}^{n} m_{ij}(zy)\varphi_j(x)$$

$$= \sum_{k,j=1}^{n} m_{ik}(z)m_{kj}(y)\varphi_j(x).$$

Since $\{\varphi_i\}$ $(1 \le i \le n)$ are linearly independent,

$$m_{ij}(zy) = \sum_{k=1}^{n} m_{ik}(z)m_{kj}(y).$$

But this shows that $M_{zy} = M_z M_y$. In other words, $z \to M_z$ is a homomorphism of G into $\mathcal{M} = \{M_x\}$. This proves that $\mathcal{M}$ is a representation of G and so $m_{ij} \in \mathfrak{F}$.

By replacing z by x and x by e in (ii), we have

$$\varphi_i(x) = \sum_{j=1}^{n} m_{ij}(x)\varphi_j(e).$$

This establishes that each φ_i is a linear combination of functions in $\mathfrak{F}$, and thus the lemma is proved.

Theorem 9 (*Peter-Weyl*). *Let G be a metrizable compact semitopological group. Let $\mathfrak{M}$ denote the set of all functions belonging to inequivalent irreducible unitary representations of G. Then $\mathfrak{M}$ is total in $C(G)$, where the latter is endowed with the sup norm topology. In other words, every continuous complex-valued function on G can be approximated uniformly by functions coming from irreducible inequivalent unitary representations of G.*

PROOF. Let $\mathcal{M}$ be any representation of G and $M_x = (m_{ij}(x)) \in \mathcal{M}$, $x \in G$. Then there exists a constant regular matrix T such that $T^{-1}M_x T = U_x$, where

$$U_x = \begin{bmatrix} U_1(x) & & & & \\ & U_2(x) & & & \text{\Large 0} \\ & & \cdot & & \\ & & & \cdot & \\ \text{\Large 0} & & & & U_m(x) \end{bmatrix},$$

and each $U_i(x)$ is an irreducible unitary matrix. This shows that each function in $\mathfrak{F}$ (Lemma 8) is a linear combination of functions from $\mathfrak{M}$. But each function in $\mathfrak{F}_0$ (Lemma 7) is a linear combination of functions from $\mathfrak{F}$ by Lemma 8. Therefore, each function in $\mathfrak{F}_0$ is a linear combination of functions from $\mathfrak{M}$. Since $\mathfrak{F}_0$ is total in $C(G)$ by Lemma 7, the proof of the theorem is thus completed.

Corollary 5. *Let $\mathfrak{R}$ denote the set of all irreducible inequivalent unitary representations of a metrizable compact semitopological group G. Then for every $x \in G$, $x \neq e$, there exists a representation $\mathcal{M} = \{M_y\}$, $y \in G$ in $\mathfrak{R}$ such that $M_x \neq I$, in which I is the identity matrix in $\mathcal{M}$.*

PROOF. Since G is completely regular, there exists a real valued continuous function f on G such that $f(x) \neq f(e)$.

Now suppose $M_x = I$ for each $\mathcal{M}$ in $\mathfrak{R}$. Then we have $m_{ij}(x) = m_{ij}(e)$ for all m_{ij}'s in $\{M_x\} = \mathcal{M}$. Let $\varepsilon = |f(x) - f(e)| > 0$. By the theorem of Peter-Weyl, there exist k functions in $\mathfrak{M}$ by a linear combination of which f can be approximated uniformly. That is, there exist m_p, $1 \leq p \leq k$, such that

$$\left| f(y) - \sum_{p=1}^{k} c_p m_p(y) \right| < \varepsilon/2$$

for all $y \in G$, where c_p are complex numbers. In particular, for $y = e$, we have

$$\left| f(e) - \sum_{p=1}^{k} c_p m_p(e) \right| < \varepsilon/2,$$

and also

$$\left| f(e) - \sum_{p=1}^{k} c_p m_p(x) \right| < \varepsilon/2,$$

because $m_p(x) = m_p(e)$ for each p by assumption. But then

$$\left| f(x) - \sum_{p=1}^{k} c_p m_p(x) \right| = \left| f(x) - f(e) + f(e) - \sum_{p=1}^{k} c_p m_p(x) \right|$$

$$\geq |f(x) - f(e)| - \left| f(e) - \sum_{p=1}^{k} c_p m_p(x) \right| \geq \varepsilon/2,$$

which is a contradiction. Thus the corollary is established.

Remark. Whenever Corollary 5 is true for a topological group, we say that the topological group has lots of irreducible representations.

Let G be a locally compact topological abelian group. Let $\mathcal{M} = \{M_x\}$ be an irreducible representation of G. Since, for $x, y \in G$, $xy = yx$ implies $M_x M_y = M_y M_x$, by Corollary 2, §38, it follows that each M_x is a matrix of order 1×1. But then if we identify 1×1 matrices with complex numbers, we see that each irreducible representation of G is a continuous homomorphism of G into the set of complex numbers. By definition, a continuous homomorphism of a topological group into the set of complex numbers is called a character. Thus, reinterpreting Corollary 5, we see that a compact metrizable semitopological abelian group has lots of characters.

It is of interest here to mention that Gelfand and Raikov have shown that all locally compact Hausdorff topological groups have lots of irreducible representations by unitary operators on a Hilbert space. See Exercise 6 for detailed remarks.

Definition 5. A complex-valued function f on a group G is said to be *invariant under inner automorphisms* if

$$f(a^{-1}xa) = f(x)$$

for all $a \in G$.

Remark. From the Remark following Corollary 4, §39, it follows that the trace function is invariant under inner automorphism.

Theorem 10. *Let G be a metrizable compact semitopological group. Let $I(G)$ denote the set of all continuous complex-valued functions invariant under inner automorphisms of G, with the topology of $I(G)$ that induced from $C(G)$ (cf. Lemma 7). Let $\{\chi\}$ denote the family of all trace functions of irreducible inequivalent unitary representations of G. Then $\{\chi\}$ is countable and total in $I(G)$.*

PROOF. By Corollaries 3 and 4, §39, $\{\chi\}$ is a set of orthonormal functions. Hence, by Theorem 6, §40, $\{\chi\}$ is countable. For the second part, let $f \in I(G)$ and let $\varepsilon > 0$ be given. By the Peter-Weyl theorem, there exists a finite family of functions m_k in $\mathfrak{M}$ (Theorem 9) such that

(i) $$|f(x) - g(x)| < \varepsilon \qquad \text{for all } x \in G,$$

in which g is the linear combination of m_k's. Owing to invariance of f under inner automorphisms,

$$\int_G f(y^{-1}xy)\, dy = \int_G f(x)\, dy = f(x).$$

Therefore, in view of (i), by putting $h(x) = \int_G g(y^{-1}xy)\, dy$, we have

$$|f(x) - h(x)| = \left| f(x) - \int_G g(y^{-1}xy)\, dy \right|$$

$$= \left| \int_G f(y^{-1}xy)\, dy - \int_G g(y^{-1}xy)\, dy \right| < \varepsilon$$

for all $x \in G$. Since g is a linear combination of functions from $\mathfrak{M}$ and since $\mathfrak{M} \subset \mathfrak{F}$ (Lemma 8), h is a linear combination of functions from $\mathfrak{F}$.

Moreover, we show that h is invariant under inner automorphisms. For,

$$h(a^{-1}xa) = \int_G g(y^{-1}a^{-1}xay)\,dy$$

$$= \int_G g((ay)^{-1}x(ay))\,dy$$

$$= \int_G g(z^{-1}zx)\,dz$$

$$= h(x).$$

Now to complete the proof of the theorem, it will suffice to show that

$$h = \sum_{i=1}^{n} \alpha_i \chi_i, \quad \text{for some } n, \quad \text{where } \chi_i \in \{\chi\}.$$

Let $h = \sum_{i=1}^{k} h^{(t)}$, in which each $h^{(t)} = \sum_{i,j=1}^{n} c_{ij} m_{ij}^{(t)}$ is composed of functions in $\mathfrak{F}$ and in which for each t, $m_{ij}^{(t)}$'s belong to one irreducible representation $\mathcal{M}^{(t)} = \{M_x^t\}$, $x \in G$. (This decomposition is possible because h is a linear combination of functions from $\mathfrak{F}$.)

We show that $h^{(t)}$ is invariant under inner automorphisms. Indeed, $h^{(t)}(a^{-1}xa)$ can be written as a linear combination of functions in $\mathcal{M}^{(t)}$. We know that any two distinct functions in $\mathcal{M}$ belonging to distinct unitary matrices or to inequivalent irreducible representations, are orthogonal (Theorems 3 and 4, §39). Since $h(a^{-1}xa) = h(x)$,

$$\sum_{t=1}^{k} h^{(t)}(a^{-1}xa) = \sum_{t=1}^{k} h^{(t)}(x)$$

must hold termwise, i.e., $h^{(t)}(a^{-1}xa) = h^{(t)}(x)$, since for each t, $m_{ij}^{(t)}$ can be chosen to be linearly independent. Hence, $h^{(t)}$ is invariant under inner automorphisms.

Now we shall show that $h^{(t)} = \alpha_t \chi_t$ for $1 \leq t \leq k$, where α_t is a complex number. Since $h^{(t)}$ is invariant under inner automorphisms, i.e.,

$$h^{(t)}(x) = h^{(t)}(a^{-1}xa),$$

we see that

$$\sum_{i,j=1}^{n} c_{ij} m_{ij}^{(t)}(x) = \sum_{i,j=1}^{n} c_{ij} m_{ij}^{(t)}(a^{-1}xa)$$

$$= \sum_{i,j,k,p=1}^{n} c_{ij} m_{ik}^{(t)}(a^{-1}) m_{kp}^{(t)}(x) m_{pj}^{(t)}(a).$$

From this it follows that

$$c_{ij} = \sum_{k,p=1}^{n} m_{ik}^{(t)}(a) c_{kp} m_{pj}^{(t)}(a^{-1}).$$

If $C = (c_{ij})$, then the above equation can be written as

$$CM_a^{(t)} = M_a^{(t)}C.$$

Since $\mathcal{M}^{(t)} = \{M_x^{(t)}\}$ is an irreducible representation, by Corollary 1, §38, $C = \alpha_t I$, α_t complex. Therefore, $c_{ij} = 0$ or α_t for $i \neq j$, or $i = j$, respectively. Hence,

$$h^{(t)} = \sum_{i=1}^{n} \alpha_t m_{ii}^{(t)} = \alpha_t \sum_{i=1}^{n} m_{ii}^{(t)} = \alpha_t \chi_t$$

for each $t (1 \leq t \leq k)$. Therefore,

$$h = \sum_{t=1}^{k} h^{(t)} = \sum_{t=1}^{k} \alpha_t \chi_t,$$

and the proof is completed.

43. STRUCTURE OF METRIZABLE COMPACT GROUPS

Recall that $\mathbf{U}_n$ (Notation 3, §27, Chapter IV) denotes the group of unitary square matrices of order n. $\mathbf{U}_n$ is a compact topological group (Theorem 11, §27, Chapter IV). Let $\mathbf{U} = \prod_{n=1}^{\infty} \mathbf{U}_n$. Then $\mathbf{U}$ is also a topological group and it is compact by Tychonoff's theorem.

Theorem 11.   *Every metrizable compact semitopological group G is homeomorphic to a closed subgroup of* $\mathbf{U}$.

PROOF.   Observe that G is a topological group. Let $\{\mathcal{M}^{(n)}\}$ denote the system of all unitary representations of G. Let d_n denote the degree of $\mathcal{M}^{(n)}$, i.e., the order of matrices in $\mathcal{M}^{(n)}$ is $d_n \times d_n$. Let $\{n_k\}$ be an increasing subsequence of natural numbers $\{n\}$ such that $n_k \geq d_k$ for each k. Then the set of unitary matrices of order d_k is isomorphic to a subgroup of $\mathbf{U}_{n_k}$ for each k. Hence, the representation $\mathcal{M}^{(d_k)}$ of G can be identified with a subgroup of $\mathbf{U}_{n_k}$. For each $x \in G$, define

$$f(x) = (M_x^{(n)}),$$

where $M_x^{(n)} \in \mathcal{M}^{(n)}$ for each $n \in \{n_k\}$, and equal to I (identity matrix) otherwise. Then $f(x) \in \mathbf{U}$.

For $x, y \in G$, we have

$$f(xy) = (M_{xy}^{(n)}) = (M_x^{(n)} M_y^{(n)}) = (M_x^{(n)})(M_y^{(n)}) = f(x)f(y),$$

because $\mathcal{M}^{(n)}$ is a representation of G for each n and therefore $M_{xy}^{(n)} = M_x^{(n)} M_y^{(n)}$. This shows that f is a homomorphism of G into $\mathbf{U}$.

To show that f is 1:1, let $x \neq e$. Then by Corollary 5, §42, there exists a positive integer k such that $M_x^{(k)} \neq I_x$. Hence, $f(x) \neq (I_x^{(n)})$.

Since the mapping: $x \to M_x^{(n)}$ is continuous for each n (because $\mathcal{M}^{(n)}$ is a representation of G), so is f. Hence, $f(G)$ is a compact (and therefore closed) subgroup of $\mathbf{U}$, and also Hausdorff. This shows that G is homeomorphic with $f(G)$.

Theorem 12. *Let G be a metrizable compact semitopological group. Then there exists a strictly decreasing sequence $\{N_n\}$ of invariant closed subgroups of G such that G/N_n is a Lie group (Definition 5, §29, Chapter IV) and $\bigcap N_n = \{e\}$.*

Proof. Let $\mathbf{U} = \prod_{n=1}^{\infty} \mathbf{U}_n$ be the product of unitary groups $\mathbf{U}_n$ of square matrices of order n. Let $\mathbf{V}_n = \prod_{k=1}^{n} \mathbf{U}_k$ and $\mathbf{W}_n = \prod_{k=n+1}^{\infty} \mathbf{U}_k$. By Proposition 15, §29, Chapter IV, $\mathbf{V}_n$ is a Lie group. Since $\mathbf{W}_n$ is an invariant closed subgroup of $\mathbf{U}$, and $\mathbf{U}/\mathbf{W}_n$ is homeomorphic with $\mathbf{V}_n$, $\mathbf{U}/\mathbf{W}_n$ is also a Lie group.

Furthermore, for each n, $\mathbf{W}_n \supset \mathbf{W}_{n+1}$ and $\bigcap_{n\geq 1} \mathbf{W}_n = \{I\}$, I identity of $\mathbf{U}$. By Theorem 11, G can be identified with a closed subgroup of $\mathbf{U}$. Let $N_n = G \cap \mathbf{W}_n$. Then N_n is an invariant closed subgroup of G because $\mathbf{W}_n$ is an invariant closed subgroup of $\mathbf{U}$, and $N_n \supset N_{n+1}$ for each n. Moreover, $\bigcap_{n\geq 1} N_n = G \cap \left(\bigcap_{n\geq 1} \mathbf{W}_n \right) = \{e\}$. Now consider the mapping $\varphi_n : \mathbf{U} \to \mathbf{U}/\mathbf{W}_n$. Let $\varphi_n(G) = G_n$ for each n. Since $\mathbf{U}/\mathbf{W}_n$ is a Lie group for each n, G_n being a compact and therefore closed subgroup of a Lie group is also a Lie group. But by the isomorphism theorem we have G_n homeomorphic with G/N_n, and thus G/N_n is a Lie group for each n. This completes the proof.

Theorem 13 (*The structure theorem*). *Every metrizable compact semitopological group G is homeomorphic with the inverse limit (§25) of a sequence of Lie groups.*

Proof. By Theorem 12, there exists a decreasing sequence $\{N_n\}$ of invariant closed subgroups of G such that $\bigcap_{n=1}^{\infty} N_n = \{e\}$ and G/N_n is a Lie group.

Let $G_n = G/N_n$ and let $\varphi_n : G \to G_n$. Since $N_{n+1} \subset N_n$ for each n, there exists a unique homomorphism f_n of G_{n+1} onto G_n such that for $x \in G$,

$$\varphi_n(x) = f_n(\varphi_{n+1}(x)).$$

Now $\{G_n, f_n\}$ is a countable inverse system of topological groups. Let G_∞ denote the inverse limit of the system. We shall show that G is homeomorphic with G_∞.

For each $x \in G$, let
$$f(x) = (\varphi_n(x)).$$

Since $\varphi_n(x) \in G_n$ for each n, $f(x) \in \prod_{n=1}^{\infty} G_n$. Moreover, for each n,

$$\varphi_n(x) = f_n(\varphi_{n+1}(x))$$

means that for each $\dot{x}_{n+1} \in G_{n+1}$, $f_n(\dot{x}_{n+1}) \in G_n$. Hence, this proves that $f(x) \in G_\infty$. Furthermore,

$$f(xy) = (\varphi_n(xy)) = (\varphi_n(x)\varphi_n(y)) = (\varphi_n(x))(\varphi_n(y)),$$

because φ_n is a homomorphism. This shows that f is a homomorphism of G into G_∞.

Since each φ_n, being the canonical mapping, is continuous and open, f is also continuous and open. We have to show that f is 1:1 and onto.

Let $(y_n) \in G_\infty$. Then $y_n \in G_n$ for each n. Therefore, there exists $x_n \in G$ such that $\varphi_n(x_n) = y_n$. Since φ_n is continuous and onto, $\varphi_n^{-1}(y_n)$ is a closed subset of G and, therefore, compact because G is compact.

Clearly, $\varphi_{n+1}^{-1}(y_{n+1}) \subset \varphi_n^{-1}(y_n)$. By compactness of G, $\bigcap\limits_{n=1}^{\infty} \varphi_n^{-1}(y_n) \neq \varnothing$. Let $x \in \bigcap\limits_{n=1}^{\infty} \varphi_n^{-1}(y_n)$. Then $\varphi_n(x) = y_n$ for each n and hence,

$$f(x) = (\varphi_n(x)) = (y_n).$$

This proves that f is onto.

To show that f is 1:1, suppose $f(x)$ is the identity in G_∞. That means $\varphi_n(x)$ is the identity in G_n for each n. But this shows that $x \in N_n$ for each n. Since $\bigcap\limits_{n \geq 1} N_n = \{e\}$, $x = e$. This completes the proof.

Remark. As pointed out earlier, the results of this chapter can be proved more generally when G is any compact Hausdorff topological group and when the representation of G is given by the continuous endomorphisms of a Hilbert space. See Exercises 5–13 below.

Exercises

1. Let $\mathcal{M}$ be a unitary representation of a compact Hausdorff topological group. Let χ denote the trace function of $\mathcal{M}$. Show that $\mathcal{M}$ is irreducible if, and only if,

$$\int_G \chi(x)\bar{\chi}(x)\, dx = 1.$$

2. Let G be a finite group. Find a total family of orthonormal functions on G.

3. Let T denote the torus group, i.e., the quotient group of the reals modulo the group of the integers.

(a) Show that $\{\varphi_n(x) = e^{2\pi inx}\}$, $i = \sqrt{-1}$, is a total orthonormal family of continuous functions on T.

(b) Show also that $\{\varphi_n(x)\}$ is the totality of all inequivalent irreducible unitary representations of T.

4. Show that a totally disconnected metrizable compact group is the inverse limit of a sequence of finite groups.

5. Let H be a complex (or real) Hilbert space. Let $B(H)$ denote the set of all linear and continuous mappings of H into itself. Each $T \in B(H)$ is called an *operator*.

(a) Show that $B(H)$ is a Banach algebra (cf. Chapter IX) with the norm $\|T\| = \sup\limits_{\|x\| \leq 1} \|T(x)\|$, $T \in B(H)$, $x \in H$. The multiplication is the composition of mappings, i.e., if T_1, $T_2 \in B(H)$ then $T_1 T_2$ is defined by $T_1 T_2(x) = T_1(T_2(x))$ for all $x \in H$. Show that $B(H)$ has a unit element I (identity mapping).

(b) If $\langle x, y \rangle$ denotes the scalar product for H, show that for each $T \in B(H)$ there exists a $T^* \in B(H)$ such that $\langle T(x), y \rangle = \langle x, T^*(y) \rangle$ for all $x, y \in H$. (Hint: Use Theorem 6, §11, Chapter I.)

(T^* is called the *adjoint* or *conjugate* of T. If $T = T^*$, then T is said to be *self-adjoint*, or *real* or *Hermitian*. T is called *positive or* $T \geq 0$ if $\langle T(x), x \rangle \geq 0$ for all $x \in H$. If for T_1, $T_2 \in B(H)$, $T_1 - T_2$ is positive, we write $T_2 \leq T_1$. T is called *unitary* if, and only if, $\langle T(x), T(y) \rangle = \langle x, y \rangle$ for all $x, y \in H$. T is called *invertible* if there exists $S \in B(H)$ such that $ST = ST = I$. If T is invertible, S is unique and is called the *inverse* of T. S is denoted by T^{-1}.)

(c) Show the following:

(i) $\|T\| = \|T^*\|$;

(ii) $T^{**} = T$;

(iii) $(T_1 + T_2)^* = T_1^* + T_2^*$;

(iv) $(\alpha T)^* = \bar{\alpha} T^*$;

(v) $(T_1 T_2)^* = T_2^* T_1^*$;

(vi) $\|T T^*\| = \|T\|^2$;

(vii) if T is invertible, so is T^* and $(T^*)^{-1} = (T^{-1})^*$;

(viii) T is Hermitian if, and only if, $\langle T(x), x \rangle$ is real for each $x \in H$;

(ix) T is unitary if, and only if, T^{-1} exists and $T^{-1} = T^*$;

(x) For S, $T \in B(H)$, if $S \geq 0$, $T \geq 0$ and $ST = TS$ then $ST \geq 0$.

(*Remark:* If H is a finite-dimensional Hilbert space then H can be identified with C^n. Hence, each $T \in B(H)$ can be identified with a square matrix of order $n \times n$. Thus, the properties of elements in $B(H)$ discussed above can be interpreted in terms of properties of matrices. For example, T^* is the conjugate transpose of the matrix T, and so on.)

(d) Let $T \in B(H)$ be Hermitian. Let $\alpha = \inf\limits_{\|x\|=1} \langle T(x), x \rangle$ and $\beta = \sup\limits_{\|x\|=1} \langle T(x), x \rangle$. Show that $\|T\| = \max(|\alpha|, |\beta|)$, by proving first that $\|T\| = \sup\limits_{\|x\|=1} |\langle T(x), x \rangle|$.

(e) *Spectral Theorem.* Let $T \in B(H)$ be Hermitian with lower and upper bounds α and β, respectively (see (d)). Then there exists a unique family $\{P_\lambda : \lambda \text{ real}\}$ of projection mappings P_λ of H such that

(i) $P_\lambda \leq P_\mu$ if $\lambda < \mu$;

(ii) P_λ is continuous from the left, i.e., $\|P_{\lambda-\varepsilon}(x) - P_\lambda(x)\| \to 0$ as $\varepsilon \to 0^+$;

(iii) $P_\lambda = 0$ if $\lambda \leq \alpha$, and $P_\lambda = I$ if $\lambda \geq \beta$;

(iv) $T \leq \lambda I$ on the range of P_λ and $T \geq \lambda I$ on the range of $I - P_\lambda$;

(v) For any $S \in B(H)$, $TS = TS$ if and only if $SP_\lambda = P_\lambda S$ for each λ;

(vi) If $\varphi \in \Phi$ (see below), then $\varphi(T) = \lim \sum_i \varphi(\lambda_i)(P_{\lambda_i} - P_{\lambda_{i-1}}) = \int \varphi(\lambda) \, dP_\lambda$,

where the limit is taken in the norm of $B(H)$ and has the same sense as in the familiar Riemann sums for the Riemann integrals of real-valued functions.

(*Indication of proof:* Let Φ^+ denote the set of all nonnegative real-valued functions on the closed bounded interval $[\alpha, \beta]$ such that each $\varphi \in \Phi^+$ is the pointwise limit of an increasing sequence of polynomials. Let $\Phi = \{\varphi : \varphi^+, \varphi^- \in \Phi^+\}$ in which $\varphi^+(t) = \varphi(t)$ or 0 according as $\varphi(t) \geq 0$ or $\varphi(t) < 0$, and $\varphi^-(t) = -\varphi(t)$ or 0 according as $\varphi(t) \leq 0$ or $\varphi(t) > 0$. By the Weierstrass approximation theorem Φ contains all continuous functions on $[\alpha, \beta]$. Also Φ contains functions

$$\varphi_\lambda(t) = \begin{cases} 1 & \text{if } t < \lambda \\ 0 & \text{if } t \geq \lambda \end{cases}.$$

Show that Φ is an algebra. For each polynomial p, the mapping $p \to p(T)$ is an algebra homomorphism of the algebra of real polynomials into the algebra $B(H)$. Using (x) of (c), show that this homomorphism is order preserving, i.e., if for two polynomials $p(t) \leq q(t)$ for real t, then $p(T) \leq q(T)$ in $B(H)$. Extend this homomorphism to Φ in the following way:

First if $\{p_n\}$ is a sequence of polynomials converging to $\varphi \in \Phi^+$ pointwise, then show that $p_n(T)$ converges to $\varphi(T)$ pointwise and this limit $\varphi(T)$ is independent of a particular choice of the sequence $\{p_n\}$ of polynomials, i.e., if $\{p_n\}$ and $\{q_n\}$ are two polynomials such that $\lim_{n\to\infty} p_n = \lim_{n\to\infty} q_n = \varphi \in \Phi$ then $\lim_{n\to\infty} p_n(T) = \lim_{n\to\infty} q_n(T)$. To pass from Φ^+ to Φ, let $\varphi \in \Phi$. Then $\varphi = \varphi_1 - \varphi_2$, φ_1, $\varphi_2 \in \Phi^+$. Put $\varphi(T) = \varphi_1(T) - \varphi_2(T)$. Finally, show that the algebra homomorphism $\varphi \to \varphi(T)$ of Φ into $B(H)$ is order preserving. Now, putting $P_\lambda = \varphi_\lambda(T)$, conclude the proof. See Riesz.[39])

6. (a) Let $T \in B(H)$. If for some real number $M > 0$, $\|T(x)\| \geq M \|x\|$ for all $x \in H$, the range $\{T(x) : x \in H\}$ of T is a closed subset of H.

(b) T is invertible if, and only if, the range of T is dense in H and if there exists a real number $M > 0$ such that $\|T(x)\| \geq M \|x\|$ for all $x \in H$.

(c) Let $T \in B(H)$. The set of complex numbers λ for which $T - \lambda I$ is not invertible, is called the *spectrum* of T and is denoted by $Sp(T)$. The set of complex numbers λ for which $T(x) = \lambda x$ for some nonzero $x \in H$, is called the *point-spectrum*. If a nonzero x is such that $T(x) = \lambda x$, then λ is called a *characteristic value* of T and x is called a *characteristic vector* corresponding to λ.

(i) The point-spectrum of an operator $T \in B(H)$ is a subset of $Sp(T)$.

(ii) If $T \in B(H)$ is Hermitian, $Sp(T)$ is a subset of the real line. (Hint: If λ is in the point-spectrum, then trivially $\lambda = \bar{\lambda}$. If $\lambda \in Sp(T)$, and λ is not in the point-spectrum, show that the range of $T - \lambda I$ is dense if λ is complex and that $\|T(x) - \lambda x\| \geq 2^{-1}|\lambda - \bar{\lambda}| \|x\|$. Use (b) to arrive at a contradiction.)

(iii) If $T \in B(H)$ is Hermitian, show that $Sp(T)$ consists of all real λ such that for all $\varepsilon > 0$, $P_{\lambda+\varepsilon} \neq P_{\lambda-\varepsilon}$, where $\{P_\lambda\}$ is the spectral resolution of T given in Exercise 5(e). (Hint: Use (b) and (c) (ii).)

(d) For each $T \in B(H)$, $Sp(T)$ is nonempty. (Hint: Use Proposition 2(d), §49, Chapter IX.)

(e) If $T \in B(H)$ is Hermitian and if λ_1, λ_2, $\lambda_1 \neq \lambda_2$, are characteristic values of T such that $T(x_1) = \lambda_1 x_1$, and $T(x_2) = \lambda_2 x_2$, then x_1 and x_2 are orthogonal, i.e., $\langle x_1, x_2 \rangle = 0$.

7. An operator T on a Banach space E is said to be *completely continuous*

if for each bounded sequence $\{x_n\}$ in E, $\{T(x_n)\}$ contains a convergent subsequence or, equivalently, the unit ball $\{x \in E : \|x\| \leq 1\}$ is mapped onto a relatively compact subset of E.

(a) Each continuous endomorphism of a finite-dimensional Banach space is completely continuous.

(b) The set of all completely continuous operators on a Banach space E is a norm-closed subalgebra of $B(E)$, the algebra of all continuous endomorphisms of E.

(c) If T is a completely continuous operator on a Banach space E, the kernel of $I - T$ is a closed finite-dimensional subspace of E. (Hint: First show that if F is a closed proper subspace of E and if $0 \leq c < 1$, then there exists $x_0 \in E$ such that $\|x_0 - y\| \geq c$ for all $y \in F$. From this derive that if E is an infinite-dimensional Banach space then there exists a sequence $\{x_n\} \subset E$ such that $\|x_n\| = 1$, $\|x_n - x_m\| > \frac{1}{2}$, $n \neq m$. If the kernel of $I - T$ is not finite-dimensional, the existence of such a sequence contradicts the complete continuity of T.)

(d) (i) Let H be a Hilbert space with $\langle , \rangle$ as scalar product function. Let $y, z \in H$ be any two elements. For each $x \in H$, define $T(x) = \langle x, y \rangle z$. Then T is a completely continuous operator on H.

(ii) More generally, if $y_i, z_i \in H$, $1 \leq i \leq n$, then the operator $T(x) = \sum_{i=1}^{n} \langle x, y_i \rangle z_i$ is completely continuous. (Hint: Use (b) and (d)(i).)

(e) If $T \in B(H)$ is a nonzero Hermitian and completely continuous operator on a Hilbert space H, then T has at least one nonzero characteristic value.

(Hint: By Exercise 5(d), there exists a sequence $\{x_n\}$, $\|x_n\| = 1$, such that $\langle T(x_n), x_n \rangle \to \lambda = \pm \|T\| \neq 0$. Since $\|(T - \lambda)x_n\|^2 = \|T(x_n)\|^2 - 2\lambda \langle T(x_n), x_n \rangle + \lambda^2$ and since, by the Schwarz inequality, $\|T(x_n)\|^2 \leq \|T\|^2 = \lambda^2$, we obtain $\limsup\limits_{n \to \infty} \|(T - \lambda)x_n\|^2 \leq 0$. Hence, $T(x_n) - \lambda x_n \to 0$. Owing to complete continuity of T, there is a convergent subsequence $\{T(x_{n_k})\}$ of $\{T(x_n)\}$. Hence, $\{\lambda x_{n_k}\}$ is convergent (because $\lambda \neq 0$) to λx_0, for some $x_0 \in H$, $\|x_0\| = 1$, and hence $T(x_0) = \lambda x_0$. Hence λ is a nonzero characteristic value.)

(f) If $T \in B(H)$ is completely continuous and Hermitian, then for $x \in H$, $$T(x) = \sum_{i=1}^{\infty} \lambda_i \langle x, x_i \rangle x_i,$$ in which $\{x_i\}$ is a countable orthonormal family of elements in H and $T(x_i) = \lambda_i x_i$ for each $i \geq 1$.

(Hint: Call λ and x_0 of the hint in (e) λ_1 and x_1 respectively and assume $\|x_1\| = 1$. For each $y \in x_1^{\perp}$, the orthogonal complement of x_1, $\langle T(y), x_1 \rangle = \langle y, T(x_1) \rangle = \lambda_1 \langle y, x_1 \rangle = 0$ implies T leaves $x_1^{\perp}$ invariant. T, regarded as an operator on $x_1^{\perp}$, is Hermitian and completely continuous. If T reduces to zero on $x_1^{\perp}$, then T is given by $T(x) = \lambda_1 \langle x, x_1 \rangle x_1$, $x \in H$. For, each $x \in H$ can be written as a sum $x = \langle x, x_1 \rangle x_1 + y$, where $\langle y, x_1 \rangle = 0$ and thus $T(x) = \langle x, x_1 \rangle T(x_1) + T(y) = \lambda_1 \langle x_1 x_1 \rangle x_1$, because $T(y) = 0$. If T does not reduce to zero on $x_1^{\perp}$, we may repeat the argument used in the hint for (e) to obtain λ_2 and x_2 such that $T(x_2) = \lambda x_2$. Thus, as before either $T(x) = \lambda_1\langle x, x_1 \rangle x_1 + \lambda_2\langle x, x_2 \rangle x_2$, or we continue to obtain $T(x) = \sum_{i=1}^{n} \lambda_i\langle x, x_i \rangle x_i$, $x \in H$. To obtain the infinite expansion $T(x) = \sum_{i=1}^{\infty} \lambda_i\langle x, x_i \rangle x_i$, we see that if $y \in M$, the subspace of H spanned by $\{x_i\}$, then by the Parséval equality $y = \sum_{i=1}^{\infty} \langle y, x_i \rangle x_i$. If $z \in M^\perp$, then show that $\langle T(z), z \rangle = 0$. By Exercise 5(d), $T(z) = 0$. Thus, if x is any element in H, then $x = y + z$, $y \in M$, $z \in M^\perp$. Hence, $T(x) = T(y) = \sum_{i=1}^{\infty} \lambda_i\langle y, x_i \rangle x_i)$.

8. Let X be a measure space (i.e., there is a σ-ring $\mathcal{B}$ of subsets of X and there is a measure m defined on $\mathcal{B}$) such that $m(X) < \infty$. Let K be a complex-valued function on $X \times X$ such that $K \in L_2(X \times X)$ (cf., §33, Chapter VI). For each $f \in L_2(X)$, put $K(f) = g$, where

$$K(f)(x) = g(x) = \int_X K(x, y)f(y) \, dm(y).$$

(Recall that $L_2(X)$ is a Hilbert space with scalar product function defined by $\langle f, g \rangle = \int_X f\bar{g}.$)

(a) K is an operator on $L_2(X)$.

(b) K is Hermitian if, and only if, $K(x, y) = \overline{K(y, x)}$.

(c) K is completely continuous.

In particular, let $X = [a, b]$, $-\infty < a < b < \infty$, and let m be the Lebesgue on $[a, b]$. Let $k(x, y)$ be a continuous real-valued function such that $k(x, y) = k(y, x)$. For each continuous real-valued function f on $[a, b]$, put

$$k(f) = \int_a^b k(x, y)f(y) \, dy.$$

Then k is a Hermitian completely continuous operator on $C([a, b])$. (Hint: To show that k is completely continuous, let $\{f_n\}$ be a bounded sequence in $C([a, b])$. Then $\{k(f_n)\}$ is uniformly bounded and equicontinuous. Apply Ascoli's theorem (§10, Chapter I).

9. *Spectral theorem for Hermitian completely continuous operators*

Let H be a separable Hilbert space and let $T \in B(H)$ be a Hermitian and completely continuous operator. Then:

(i) H is spanned by the characteristic vectors or functions of T.

(ii) For each $\varepsilon > 0$ there exists only a finite number of characteristic values λ such that $|\lambda| > \varepsilon$. (This shows that the characteristic values tend to zero and that there is only a countable number of characteristic values.)

(iii) For each $\lambda \neq 0$, the vector space V_λ, consisting of $x \in H$ such that $(I - \lambda T)x = 0$, is a finite-dimensional closed subspace of H. (Hint: Utilize 7(c).)

(iv) For any $S \in B(H)$, $ST = TS$ if, and only if, S leaves each V_λ invariant (i.e., $S(v) \in V_\lambda$ for each $v \in V_\lambda$).

(*Indication of proof*: Most of (i) to (iv) are derived from the following: For each $\alpha > 0$ the ranges of $P_{-\alpha}$ and $I - P_\alpha$ are finite-dimensional, where $\{P_\lambda\}$ is the family of projections given by Exercise 5(e). Since on the range of $P_{-\alpha}$, $T \leq -\alpha I$ and on the range of $I - P_\alpha$, $T \geq \alpha I$, we obtain $\|T(x)\| \geq \alpha \|x\|$ on the ranges of both $P_{-\alpha}$ and $I - P_\alpha$. Now apply the argument used in the hint for Exercise 7(c). Also use Exercise 7(d), (f).)

(As usual, define a *representation* of a locally compact Hausdorff topological group G to be a continuous homomorphism f of G into nonzero elements of $B(H)$. A representation f of G is said to be *irreducible* if the subset $f(G)$ of $B(H)$ satisfies the following property: There exists no closed proper subspace H_1 (i.e., $H_1 \neq H$ or $\{0\}$) of H such that $T(H_1) \subset H_1$ for each $T \in f(G)$. Otherwise, f is called *reducible*. If there exists a closed proper subspace H_1 of H such that $T(H_1) \subset H_1$ and $T(H_1^\perp) \subset H_1^\perp$ for all $T \in f(G)$, then f is said to be *completely reducible*. f is *unitary* if, and only if, $f(G)$ is a set of unitary elements in $B(H)$. Let f_1 be a representation of G into $B(H_1)$ and f_2 a representation of G into $B(H_2)$, where H_1 and H_2 are Hilbert spaces. Then f_1 is said to be *equivalent* to f_2 if there exists an invertible linear continuous transformation of H_1 onto H_2 such that $f_1(x) = T^{-1}f_2(x)T$ for all $x \in G$.

10. (a) Let G be a locally compact Hausdorff topological group and let f be a unitary irreducible representation of G into $B(H)$. Show that if for any $T \in B(H)$, $TU_x = U_x T$, in which $U_x \in f(G)$, $x \in G$, then $T = \alpha I$ for a complex α. (Hint: Use 5(e).)

(*Remark:* Gelfand-Raikov's theorem says that every locally compact Hausdorff topological group has lots of unitary irreducible representations. We refer the reader for a proof to Hewitt and Ross,[19] p. 343.)

(b) Every irreducible unitary representation of a compact Hausdorff topological group in $B(H)$ is finite-dimensional i.e., H is a finite-dimensional Hilbert space.

(PROOF: Let $\{U_x\}_{x \in G}$ denote an irreducible set of unitary operators representing the compact group G on H. Let $h_1, h_2, h_3 \in H$. For fixed h_3, consider

$$F(h_1, h_2) = \int_G \langle U_x h_3, h_2 \rangle \overline{\langle U_x h_3, h_1 \rangle} \, dx.$$

Then F is linear in h, $F(h_1, h_2) = \overline{F(h_2, h_1)}$ and $|F(h_1, h_2)| \leq \|h_3\|^2 \|h_1\| \|h_2\|$. By the Riesz representation theorem, $F(h_1, h_2) = \langle T_{h_3}(h_1), h_2 \rangle$ for some $T_{h_3} \in B(H)$. By the left-invariance of Haar measure, for each $U_y \in \{U_x\}$, $y \in G$

$$\langle T_{h_3} U_y(h_1), h_2 \rangle = \int_G \langle U_x h_3, h_2 \rangle \overline{\langle U_x h_3, U_y h_1 \rangle} \, dx$$

$$= \int_G \langle U_{y^{-1}x} h_3, U_{y^{-1}} h_2 \rangle \overline{\langle U_{y^{-1}x} h_3, h_1 \rangle} \, dx \quad \text{(Since } U_{y^{-1}} \text{ is unitary)}$$

$$= \int_G \langle U_x h_3, U_{y^{-1}} h_2 \rangle \overline{\langle U_x h_3, h_1 \rangle} \, dx$$

$$= \langle T_{h_3}(h_1), U_{y^{-1}} h_2 \rangle = \langle U_y T_{h_3}(h_1), h_2 \rangle.$$

Hence, $T_{h_3} U_y = U_y T_{h_3}$. By (a), $T_{h_3} = \alpha(h_3)I$, in which $\alpha(h_3)$ is a complex number. Thus, $\alpha(h_3)\langle h_1, h_2 \rangle = F(h_1, h_2)$ and, in particular,

(1) $$F(h_1, h_1) = \alpha(h_3) \|h_1\|^2$$

for all $h_1, h_3 \in H$. Interchanging h_1 and h_3 in (1) and using unimodularity of G (since G is compact), we find

$$\alpha(h_1) \|h_3\|^2 = \int_G |\langle U_x h_1, h_2 \rangle|^2 \, dx = \int_G |\langle h_2, U_x h_1 \rangle|^2 \, dx$$

$$= \int_G |\langle U_{x^{-1}} h_2, h_1 \rangle|^2 \, dx = \int_G |\langle U_x h_2, h_1 \rangle|^2 \, dx$$

$$= \alpha(h_3) \|h_1\|^2.$$

Therefore, for some constant β, $\alpha(h_1) = \beta \|h_1\|^2$. If $h_1 = h_3$ and $\|h_1\| = 1$, from (1) we obtain

$$\int_G |\langle U_x h_1, h_1 \rangle|^2 \, dx = \alpha(h_1) \|h_1\|^2 = \beta \|h_1\|^4 = \beta.$$

Since the continuous function $x \to |\langle U_x h_1, h_1 \rangle|^2$ attains its value 1 at $x = e$, $\beta > 0$.

Now let $\{\varphi_i\}$, $1 \leq i \leq n$, be an orthonormal family of vectors in H. Putting $h_1 = \varphi_1$, and $\varphi_i = h_3$ ($1 \leq i \leq n$) in (1), we obtain,

$$\int_G |\langle U_x \varphi_i, \varphi_1 \rangle|^2 \, dx = \alpha(\varphi_i) \|\varphi_1\|^2 = \beta.$$

Summing over i and observing that $\{U_x \varphi_i\}$ are orthonormal (since U_x is unitary), by Bessel's inequality

$$n\beta = \sum_{i=1}^{n} \int_G |\langle U_x \varphi_i, \varphi_1 \rangle|^2 \, dx = \int_G \sum |\langle U_x \varphi_i, \varphi_1 \rangle|^2 \, dx$$

$$\leq \int_G \|\varphi_1\|^2 = 1.$$

In other words, $n \leq \beta^{-1}$, which proves that H is finite-dimensional.)

11. Let H be a Hilbert space with scalar product $\langle \, , \rangle$ and let f be a representation of a compact Hausdorff topological group G into $B(H)$.

(a) For any fixed pair $h_1, h_2 \in H$, define

$$F(x) = \langle f(x)(h_1), f(x)(h_2) \rangle$$

for each $x \in G$. (Observe that $f(x) \in B(H)$ and, hence, $f(x)(h_1), f(x)(h_2) \in H$.) Show that F is a continuous complex-valued function on G.

(b) Every representation f of G into $B(H)$ is equivalent to a unitary representation. (Hint: Consider $K(h_1, h_2) = \int_G \langle f(x)(h_1), f(x)(h_2) \rangle \, dx$. Then K is linear in h_1 and $K(h_1, h_2) = \overline{K(h_2, h_1)}$. Using the Riesz representation theorem, show that $K(h_1, h_2) = \langle T_1(h_1), h_2 \rangle$, in which $T_1 \in B(H)$. Also $T_1 > 0$ i.e., $\langle T_1(h), h \rangle > 0$ for all $h \in H$, since $\langle T_1(h), h \rangle = \int_G \| f(x)(h) \|^2 \, dx > 0$. But then there exists $T \in B(H)$, $T \geq 0$ such that $T^2 = T_1$ and T is invertible. If $g(x) = T^{-1} f(x) T$, then g is a unitary representation.)

(c) Every representation of G into $B(H)$ is completely reducible. (Hint: Use (b) and see that if a unitary operator leaves a subspace of H invariant, then it leaves its orthogonal complement also invariant.)

(d) Every irreducible unitary representation of a compact Hausdorff abelian topological group is one-dimensional. (Hint: Use 10(a).)

(Let f be a representation of a compact Hausdorff topological group in $B(H)$, in which H is a finite-dimensional Hilbert space. Let H' denote the dual of H. The complex-valued function $F(x) = \langle f(x)(h), h' \rangle$, $h \in H$, $h' \in H'$, $x \in G$, is called a *representation function*. If $\{h_i\}(1 \leq i \leq n)$ is a basis of H and h'_j ($1 \leq j \leq n$) a basis of H', then $F_{ij}(x) = \langle f(x)h_i, h'_j \rangle$ are just the functions obtained from the matrix representations.)

12. *Schur's lemma and orthogonality relations*

(a) Let H_1 and H_2 be two Hilbert spaces and let $\mathscr{L}_1$ and $\mathscr{L}_2$ be two irreducible families of operators on H_1 and H_2, respectively. Let T be a linear mapping of H_1 into H_2 such that $T\mathscr{L}_1 = \mathscr{L}_2 T$ in the same sense as in Schur's lemma (Theorem 1, §38). Then either $T = 0$ or is an isomorphism of H_1 onto H_2. Hence, H_1 and H_2 have the same dimensions.

(b) Prove the analogue of Corollary 1, §38, on Hilbert spaces.

(c) If f and g are two irreducible inequivalent unitary representations of a compact Hausdorff topological group G, then all representation functions of f are orthogonal to all representation functions of g. (Hint: Let H_1 and H_2 be two Hilbert spaces associated with representations f and g, respectively. If $F(x) = \langle f(x)(h_1), h'_1 \rangle$ and $G(x) = \langle g(x)(h_2), h'_2 \rangle$, $x \in G$, $h_1 \in H_1$, $h'_1 \in H'_1$, $h_2 \in H_2$, $h'_2 \in H'$, putting $K(h_1, h_2) = \int_G \langle f(x)(h_1), h'_1 \rangle \overline{\langle g(x)(h_2), h'_2 \rangle} \, dx$, show that K is linear in h_1 and $K(h_1, h_2) = \overline{K(h_2, h_1)}$. Apply the Riesz representation theorem to obtain $K(h_1, h_2) = \langle T(h_1), h_2 \rangle$, for some linear mapping T of H_1 into H_2. Owing to the left invariance of the Haar integral, $K(h_1, h_2) = K(f(y)(h_1), g(y)(h_2))$ and, hence, $\langle Tf(y)(h_1), g(y)(h_2) \rangle = \langle T(h_1), h_2 \rangle$. Replacing h_2 by $g(y^{-1})h_2$, we obtain $\langle Tf(y)(h_1), h_2 \rangle = \langle g(y)T(h_1), h_2 \rangle$, since f and g are unitary. Therefore, $Tf(y) = g(y)T$. Apply (b) to conclude the proof.)

(d) Let f be an irreducible representation of a compact Hausdorff topological group G into $B(H)$, where H is a Hilbert space of dimension n and $\langle , \rangle$ its scalar product function. Let F_1 and F_2 be any two representation functions of f, then

$$\int_G F_1(x) \bar{F}_2(x) \, dx = \frac{1}{n} \langle h_1, h_2 \rangle \overline{\langle h'_1, h'_2 \rangle},$$

in which $F_1(x) = \langle f(x)(h_1), h'_1 \rangle$, $F_2(x) = \langle f(x)(h_2), h'_2 \rangle$, $h_1, h_2 \in H$ and $h'_1, h'_2 \in H'$. (Hint: Proceed as in (c), where $f = g$, and apply (b) to obtain

$T = cI$, in which c is a complex number. Using the terms of (c), we have

$$(*) \qquad c\langle h_1, h_2 \rangle = \int_G \langle f(x)(h_1), h_1' \rangle \overline{\langle f(x)(h_2), h_2' \rangle}\, dx,$$

in which $h_1, h_2 \in H$ and $h_1', h_2' \in H'$. To evaluate c, choose an orthonormal basis e_i $(1 \le i \le n)$ of H. Then from (*), by replacing h_1 and h_2 by e_i and summing over i, we obtain

$$nc = \int_G \sum_i \langle f(x)e_i, h_1' \rangle \overline{\langle f(x)e_i, h_2' \rangle}\, dx$$

$$= \int_G \left\langle \sum_i \langle h_2', f(x)e_i \rangle f(x)e_i, h_1' \right\rangle dx.$$

Since f is unitary, $\sum_i \langle h_2', f(x)e_i \rangle f(x)e_i = h_2'$. Hence, $nc = \overline{\langle h_1', h_2' \rangle}$ and conclude the proof.)

13. *The Peter-Weyl theorem*

(a) Let G be a compact Hausdorff topological group and let $C(G)$, as usual, denote the space of all complex-valued continuous functions on G, where $C(G)$ is endowed with the sup norm topology. Let $\mathfrak{R}$ denote the set of all representation functions coming from irreducible unitary representations of G in $B(H)$, where H is a Hilbert space. Then $\mathfrak{R}$ is total in $C(G)$.

(b) The same as (a) but without the word "irreducible."

(c) For each $x \in G$, $x \ne e$, there exists an irreducible unitary representation f of a compact Hausdorff topological group G into $B(H)$ such that $f(x) \ne I$, the identity of the group $f(G)$.

(d) For each neighborhood U of e there exists a closed invariant subgroup G_0 (contained in U) of the compact Hausdorff group G such that the compact quotient group G/G_0 is homeomorphic with a closed subgroup of U_n, the group of unitary square matrices of order $n \times n$, where n is the dimension of the Hilbert space H.

(e) Show that (a)$\Leftrightarrow$(b)$\Leftrightarrow$(c)$\Leftrightarrow$(d).

(Indication of proof of (a): We use results of §47, Chapter VIII. First, we observe that $L_1(G)$ is a Banach algebra (Corollary 1, §47, Chapter VIII) with convolution $\left(f * g(x) = \int_G f(xy^{-1})g(y)\, dy \right)$ as algebra multiplication. We show that the algebra multiplication in $L_1(G)$ can be extended onto $L_2(G)$ to make $L_2(G)$ an algebra. It is of interest to note that if G is a noncompact locally compact group, $L_2(G)$ cannot be an algebra. In that case, the central object of our interest would naturally be $L_1(G)$, which is a Banach algebra. However, for compact groups, we prefer $L_2(G)$ for the reasons that $L_2(G)$ in this case is a Hilbert algebra, i.e., $L_2(G)$ is a Hilbert space as well as a Banach algebra and, therefore, well-suited for our considerations.

(i) To extend the convolution for pairs of elements in $L_2(G)$, we observe that $C(G)$ is dense in $L_2(G)$ and, hence, $L_1(G)$ is dense in $L_2(G)$ because G is compact and $C(G) \subset L_1(G)$. Let $f, g \in L_2(G)$ and let $\{f_n\}$, $\{g_n\}$ be two sequences in $L_1(G)$ and converging to f and g, respectively, in the L_2-norm. If f, $g \in L_1(G)$, then $f * g \in L_1(G)$ by Proposition 4(c), §47, Chapter VIII. Clearly, $f_n * g_n \in L_1$ and it is easy to see that $\|f_n * g_n(x)\| \le \|f_n\|_1 \|g_n\|_1$ (see also (iv)). Hence, $\lim_{n \to \infty} f_n * g_n$ (in the sense of L_2-norm) exists and we put $f * g = \lim_{n \to \infty} f_n * g_n$ (in the L_2-norm). One can show that $f * g$ is independent of the choice of particular sequences $\{f_n\}$ and $\{g_n\}$ in $L_1(G)$, and $\|f * g\|_2 \le \|f\|_2 \|g\|_2$. Thus, $L_2(G)$ is a Banach algebra.

(ii) Since $L_1(G)$ has approximate identities (Proposition 5, §47, Chapter VIII), by the extension of convolution as indicated in the preceding paragraph $L_2(G)$ also has approximate identities.

(iii) For a fixed $f \in L_2(G)$, define the *left* and *right multiplications* L_f, R_f on $L_2(G)$ as follows: $L_f(g) = f * g$ and $R_f(g) = g * f$ for all $g \in L_2(G)$. If $\tilde{f} = \overline{f(x^{-1})}$, then $\tilde{L}_f = L_{\tilde{f}}$, in which $\tilde{L}_f(g) = \overline{(f * \bar{g})}$. Similarly, $\tilde{R}_f = R_{\tilde{f}}$. Furthermore, $L_f R_g = R_g L_f$ for f, $g \in L_2(G)$, which is nothing but the associative law of multiplication in the algebra $L_2(G)$ and which holds in $L_2(G)$ by the extension, since it holds in $L_1(G)$ (Proposition 4(d), §47, Chapter VIII). Since L_f and R_f are given by integral equations, L_f and R_f are completely continuous operators on a Hilbert space $L_2(G)$. (See Exercise 8.)

(iv) If f, $g \in L_2(G)$, then $f * g \in L_1(G)$ and $|f * g(x)| \le \|f\|_2 \|g\|_2$ for all $x \in G$. (Hence, if f_n, g_n, f, $g \in L_2(G)$ and $\|f_n - f\|_2 \to 0$, $\|g_n - g\|_2 \to 0$, then $\|f_n * g_n - f * g\|_\infty \to 0$ as $n \to \infty$.) For the first statement, let $\{f_n\}$, $\{g_n\}$ be in $L_1(G)$ such that $\|f_n - f\|_2 \to 0$ and $\|g_n - g\|_2 \to 0$. Then by the Schwarz inequality,

$$|f_n * g_n(x) - f_m * g_m(x)| \le \|f_n\|_1 \|g_n - g_m\|_1 + \|g_m\|_1 \|f_n - f_m\|_1.$$

Hence, $\{f_n * g_n\}$ converges uniformly to $h \in L_1(G)$. Whence, $f * g = h \in L_1(G)$. The inequality $\|f * g(x)\| \le \|f\|_2 \|g\|_2$ is obvious if $f, g \in L_1(G)$ (Proposition 4(c), §47, Chapter VIII). If f, $g \in L_2(G)$, and if the sequences $\{f_n\}$, $\{g_n\}$ in $L_1(G)$ are such that $f_n \to f$ and $g_n \to g$ in L_2-norm, then, by the first statement, $f_n * g_n(x) \to f * g(x)$ implies the required inequality.

(v) The left and right regular representations of G. Let $f \in L_2(G)$. For each $x \in G$, let $L_x : f \to f_x$ and $R_x : f \to f^x$, in which $f_x(y) = f(xy)$ and $f^x(y) = f(yx)$ for all $y \in G$. (It is clear that f_x, $f^x \in L_2(G)$ if $f \in L_2(G)$.) For each $x \in G$, L_x and R_x are continuous endomorphisms of a Hilbert space $L_2(G) = H$, say. Hence, L_x, $R_x \in B(H)$. Also, $L_{xy} = L_x \circ L_y$ and $R_{xy} = R_x \circ R_y$, composition of operators on H. Since, for $f \in L_1(G)$, the mapping $x \to f_x$ of G into $L_1(G)$ is uniformly continuous (see the proof of part (b) of Proposition 4, §47, Chapter VIII), by the extension of algebra multiplication (see (i)), $x \to f_x$ is also uniformly continuous if $f \in L_2(G)$. Hence, the mapping $x \to L_x$ (or R_x) is a continuous homomorphism of G into $B(H)$, $H = L_2(G)$. $x \to L_x$ and $x \to R_x$ are called the *left* and *right regular representations* of G. If $f \in L_2(G)$, then $L_f R_x = R_x L_f$ and $R_f L_x = L_x R_f$. Furthermore, if $f \in L_2(G)$, then $\tilde{f} \in L_2(G)$.

(vi) Let f be a self-adjoint (i.e., $f = \tilde{f}$) element in $L_2(G)$. Let $\{\lambda_i\}$ denote the nonzero characteristic values of the operator L_f on $L_2(G)$, and let H_i denote the vector space spanned by characteristic functions belonging to λ_i. Then L_f is completely continuous (see (iii)) and Hermitian (because $f = \tilde{f}$ implies $\tilde{L}_f = L_{\tilde{f}}$), each H_i is finite-dimensional by Exercise 9(iii), and R_x and R_f leave H_i invariant.

Let $R|H_i$ denote the restriction of the right regular representation $R: x \to R_x$ of G onto H_i. $R|H_i$ is a finite-dimensional representation of G. Since for $h \in H_i$, $L_f(h) = \lambda_i h$, $\lambda_i \neq 0$, or $\lambda_i^{-1}(f * h) = h$, by (iv), $h \in L_1(G)$. Furthermore, for $f_1, f_2 \in L_2(G)$, since $\langle R_x(f_1), f_2 \rangle$ is a continuous function on G, it follows that $\langle R_x|H_i(f_1), f_2 \rangle$ is a continuous function on G. Now to show that each $\varphi \in H_i$ is a representation function of $R|H_i$, let $\{\varphi_{ji}\}(1 \leq j \leq n_i)$ be an orthonormal basis of H_i. Then for $\varphi \in H_i$, we have

$$R_x(\varphi) = \sum_{j=1}^{n_i} \langle R_x(\varphi), \varphi_{ji} \rangle \varphi_{ji}.$$

In particular,

$$\varphi(x) = R_x(\varphi)(e) = \sum_{j=1}^{n_i} \langle R_x(\varphi), \varphi_{ji} \rangle \varphi_{ji}(e) = \langle R_x(\varphi), \sum_{j=1}^{n_i} \varphi_{ji}(e) \varphi_{ji} \rangle$$

which is a representation function of $R|H_i$, since $\sum_{j=1}^{n_i} \varphi_{ji}(e) \varphi_{ji} \in H_i$.

By Exercise 5(e) (vi), $L_f(g) = \int \lambda \, dP_\lambda$, because L_f is a Hermitian operator. This shows that $\|L_f(g) - \sum_{i=1}^{n} \lambda_i \varphi_i\| \to 0$ as $n \to \infty$, where $\varphi_i \in H_i$. In other words, for a fixed $f \in L_2(G)$ and $g \in L_2(G)$, each $f * g = L_f(g)$ can be approximated by linear combinations of representation functions coming from H_i. Since $L_2(G)$ has approximate identities (see (ii)), each $f \in L_2(G)$ can be approximated by $f * e_u$, where $\{e_u\}$ is the family of approximate identities. Combining the two approximations, we conclude the proof of (a).)

VIII

Duality Theory
and
Some
of
Its Applications

In this chapter, we shall discuss the concept of a dual group. We shall consider only locally compact abelian topological groups. We shall establish that the dual group of a locally compact Hausdorff abelian topological group is also of the same type, and that the compact and discrete abelian groups are duals of each other. For applications of duality theory, we shall establish some well-known theorems of harmonic analysis, e.g., the Inversion theorem and the Plancherel theorem.

44. THE CONCEPT AND TOPOLOGIES OF DUAL GROUPS

Definition 1. Let G be an abelian (additive) topological group. Let T be the one-dimensional circle group (Example 4, §19, Chapter III).

(a) A continuous homomorphism of G into T is called a *character* of G.

(b) The set of all characters of G is called the *dual group* of G and is denoted by G'. The addition in G' is defined as: For $x', y' \in G', x' + y'(x) = x'(x) + y'(x)$ for each $x \in G$. The inverse $-x'$ of x' is given by $(-x')(x) = -[x'(x)]$ for each $x \in G$.

It is elementary to show that G' is an additive abelian group. The homomorphism which maps each element of G into the identity of T is called the *identity* of G' and is denoted by $0'$. $0'$ is also called a *zero-character*.

The elements of G will be denoted by x, $y \cdots$ and those of G' by primed letters x', $y' \ldots$, etc. We shall let $\langle x, x' \rangle$ denote the *value* of the character $x' \in G'$ at $x \in G$. Clearly, $\langle x, x' \rangle \in T$.

Proposition 1. *Let G be an additive abelian group and G' its dual group. Then for x, $y \in G$ and $x' \in G'$,*

(a) $\langle x + y, x' \rangle = \langle x, x' \rangle + \langle y, x' \rangle$,
(b) $\langle x, x' + y' \rangle = \langle x, x' \rangle + \langle x, y' \rangle$,
(c) $\langle 0, x' \rangle = 0$,
(d) $\langle -x, x' \rangle = \langle x, -x' \rangle = -\langle x, x' \rangle$.

PROOF. (a) follows because x' is a homomorphism. (b) follows by the definition of addition of two homomorphisms. For (c), we note that each homomorphism of a group into another group maps the identity onto the identity. For (d), in view of (c) and (b), we note

$$0 = \langle 0, x' \rangle = \langle x - x, x' \rangle = \langle x, x' \rangle + \langle -x, x' \rangle.$$

Therefore, $\qquad\qquad\qquad \langle -x, x' \rangle = -\langle x, x' \rangle.$

By the definition of $0'$,

$$0 = \langle x, 0' \rangle = \langle x, x' - x' \rangle = \langle x, x' \rangle + \langle x, -x' \rangle,$$

and so $\qquad\qquad\qquad\qquad \langle x, -x' \rangle = -\langle x, x' \rangle.$

Thus (d) is established.

Since each $x' \in G'$ is a continuous mapping of G into T, $x' \in C(G, T) \subset T^G$. In §16, Chapter II, we have defined $\mathfrak{S}$-topologies on T^G. Thus we can endow G' with the relative $\mathfrak{S}$-topologies.

If $\mathfrak{S}$ consists of all finite subsets of G, then the $\mathfrak{S}$-topology is called the *point-open topology* or the *topology* of *simple convergence* and it is denoted by p. It is quite clear that the topology of simple convergence is the relative product topology induced from T^G.

If $\mathfrak{S}$ consists of all compact subsets of G, then the $\mathfrak{S}$-topology is called the *compact-open topology* or the *topology of uniform convergence on compact subsets*, and it is denoted by k.

Proposition 2. *G', endowed with p or k, is a Hausdorff topological group.*

PROOF. The facts that G' is a group and also a topological space have already been dealt with. What remains to be shown is that the mappings:

$$(x', y') \to x' + y' \text{ of } G' \times G' \text{ onto } G'$$

$$x' \to -x' \text{ of } G' \text{ onto } G'$$

are continuous. For this it is sufficient to show that there exists a fundamental system of neighborhoods (of $0' \in G'$) satisfying the conditions of Theorem 3, §20, Chapter III.

Let $\mathfrak{S}$ denote the family of all compact or all finite subsets of G. Let $\{V\}$ denote a fundamental system of open symmetric neighborhoods (of 0 in T) satisfying the conditions of Theorem 3, §20, Chapter III. Then the family $\{T(M, V)\}$, in which M runs over $\mathfrak{S}$ and V over $\{V\}$, forms a basis of neighborhoods of $0'$ in G'. For, let $T(M_1, V_1)$ and $T(M_2 V_2)$ be any two members in $\{T(M, V)\}$. Then there exists a $V_3 \in \{V\}$ such that $V_3 \subset V_1 \cap V_2$. Since $M_1 \cup M_2 \in \mathfrak{S}$ because the union of two compact or finite subsets of G is, respectively, compact or finite, we have, for each $x' \in T(M_1 \cup M_2, V_3)$,

$$\langle M_1 \cup M_2, x' \rangle \subset V_3 \subset V_1 \cap V_2.$$

This implies that $\langle M_1, x' \rangle \subset V_1$ and $\langle M_2, x' \rangle \subset V_2$. This shows that

$$T(M_1, V_1) \cap T(M_2, V_2) \supset T(M_1 \cup M_2, V_3).$$

Since $T(M_1 \cup M_2, V_3) \in \{T(M, V)\}$, it follows that the latter family forms a base of the neighborhoods of $0'$ in G' under p or k according as $\mathfrak{S}$ is the family of all finite or compact subsets of G.

Since for each $x' \in T(M, V)$ we have

$$\langle M, x' \rangle \subset V,$$

by the symmetry of V, it follows that

$$-T(M, V) = T(M, -V) = T(M, V).$$

Hence, each $T(M, V)$ is symmetric. This implies that the mapping: $x' \to -x'$ of G' onto G' is continuous. Furthermore, for each $V \in \{V\}$ there exists a $U \in \{V\}$ such that $U + U \subset V$. This implies that, for each $M \in \mathfrak{S}$,

$$T(M, U) + T(M, U) \subset T(M, V).$$

This proves that the mapping: $(x', y') \to x' + y'$ of $G' \times G'$ onto G' is continuous.

Since T is Hausdorff and $\mathfrak{S}$ is a covering of G, it is easily checked that p and k both are Hausdorff topologies.

45. DUAL GROUPS OF LOCALLY COMPACT ABELIAN GROUPS

Before we prove the main theorems of this section, we discuss the equicontinuous (§10, Chapter I) sets of G'. First of all we have the following:

Proposition 3. *The set $H(G, T)$ of all homomorphisms of a group G into T is a p-closed subset of T^G.*

PROOF. Clearly $H(G, T) \subset T^G$, and the former is endowed with the relative point-open topology induced from T^G. Since the projection $T^G \to T$ is continuous, so is its restriction on $H(G, T)$. Now, for any $x, y \in G$, consider

$$f(x + y) - f(x) - f(y) = 0.$$

Since $\{0\}$ is a closed subset of T, the set M_{xy} of $f \in T^G$ satisfying the above equation is p-closed in T^G owing to continuity of the projection. Clearly, $H(G, T) = \bigcap_{x,y \in G} M_{xy}$, and therefore $H(G, T)$ is p-closed.

Theorem 1. *If G is a locally compact Hausdorff abelian topological group so is G'^k.*

PROOF. We have already shown (Proposition 2, §44) that G'^k is a Hausdorff topological group. It is clear that G' is abelian. Thus, we have to show only that it is locally compact.

Let U be a neighborhood of 0 in G such that $\bar{U}$ is compact. Let V_m be an open neighborhood of 0 in T. Let us assume $V_m = \left\{ x \in T: -\dfrac{1}{m} < x < \dfrac{1}{m}, m(>0) \text{ integer} \right\}$ (cf., Exercise 12(a), Chapter III). Let

$$(1.1) \qquad T_m = \{x' \in G' : \langle x, x' \rangle \in V_m \qquad \text{for all } x \in \bar{U}\}.$$

By the definition of the topology k, T_m is a neighborhood of $0'$ in G'. Actually, it is a member of the subbase of the k-neighborhood system at $0'$ in G'. We shall show that $Cl_k T_m$ is compact. For this, we first show that T_m is equicontinuous.

Let $\varepsilon > 0$ be given. We wish to show that there exists a neighborhood U_1 of 0 in G such that, for all $x' \in T_m$ and $x, y \in U_1$, $x - y \in U_1$ implies

$$(1.2) \qquad |\langle x - y, x' \rangle| = |\langle x, x' \rangle - \langle y, x' \rangle| < \varepsilon.$$

Suppose there is no such U_1. Let n be a sufficiently large positive integer such that $\dfrac{1}{n} < \varepsilon$. Let W be a neighborhood of 0 in G such that

$$(1.3) \qquad \sum_{i=1}^{n} W_i \subset U, \qquad W_i = W$$

for all $i = 1, \ldots, n$ (Proposition 2, §20, Chapter III). By assumption then for each pair $x, y \in W$ and an $x' \in T_m$, $x - y \in W$ implies

$$(1.4) \qquad |\langle x - y, x' \rangle| \geq \varepsilon.$$

From (1.1) and (1.3) it follows that

$$j\langle x - y, x' \rangle = \langle j(x - y), x' \rangle \in V_m$$

for all $j = 1, \ldots, n$. Hence,

$$|\langle n(x - y), x' \rangle| < \frac{1}{m},$$

by the definition of V_m. But on the other hand, from (1.4) it follows that

$$|\langle n(x - y), x' \rangle| = |n \langle x - y, x' \rangle| \geq n\varepsilon > 1 > \frac{1}{m},$$

which is a contradiction because $x' \in T_m$, $x - y \in W$, and $\sum_{i=1}^{n} W_i \subset U \subset \bar{U}$, in view of the definition of T_m (1.1). Thus, the existence of a U_1 such that for all $x, y \in U_1$, $x - y \in U_1$ implies (1.2), has been established. This means that T_m is equicontinuous. Therefore, $Cl_p T_m$ in $C(G, T)$ is equicontinuous (Remark before Theorem 3, §10, Chapter I). Since $p \subset k$, $Cl_k T_m \subset Cl_p T_m$ and therefore $Cl_k T_m$, as a subset of an equicontinuous set, is also equicontinuous. Since T is compact, for each $x \in G$ and any subset H of G', $H(x) = \{f(x) : f \in H\}$ is relatively compact in T. Hence, in particular, $Cl_k T_m(x)$ is relatively compact in T for each $x \in G$. Therefore, by Ascoli's Theorem (Theorem 3, §10, Chapter I), $Cl_k T_m$ is k-compact. This proves that $0'$ has a compact k-neighborhood and so does each $x' \in G'$, since translations are homeomorphisms in a topological group. This proves that G'^k is locally compact.

Theorem 2. *If G is a locally compact abelian Hausdorff topological group satisfying the second axiom of countability, so is G'^k.*

PROOF. By Theorem 1, G'^k is, indeed, a locally compact abelian Hausdorff topological group. To prove the theorem, it is sufficient to show that G'^k is metrizable and separable.

(a) G'^k is metrizable: Since T is metrizable, let $\{V_m\}$ denote a countable fundamental system of open neighborhoods (of 0 in T) satisfying the conditions of Theorem 3, §20, Chapter III. Since G is locally compact and satisfies the second axiom of countability, there exists a countable base $\{U_n\}$ of open sets in G such that $\bar{U}_n$ is compact for each $n \geq 1$. Let

$$T\left(\bigcup_{i=1}^{n} \bar{U}_i, V_m \right) = \left\{ x' \in G' : \langle x, x' \rangle \in V_m \quad \text{and} \quad x \in \bigcup_{i=1}^{n} \bar{U}_k \right\}.$$

Then by definition, $\mathcal{U}' = \left\{ T\left(\bigcup_{i=1}^{n} \bar{U}_i, V_m \right) \right\}$ is a family of members of the base of the k-neighborhood system at $0'$ in G'. Since n and m run over all positive integers, $\mathcal{U}'$ is countable. We shall show that $\mathcal{U}'$ is a fundamental system of k-neighborhoods of $0'$ in G'. For this let $T(A, V)$ be any k-neighborhood of $0'$ in G', where A is a compact subset of G and V an open neighborhood of 0 in T. Since $\{V_m\}$ is a fundamental system of neighborhoods of 0 in T, there exists an integer m such that $V_m \subset V$. Since A is compact and $\{U_n\}$ is an open covering of G, only a finite number of U_n's, say, $\{U_{n_i}\}(1 \leq i \leq j)$, cover A. Clearly, for sufficiently large n,

$$A \subset \bigcup_{i=1}^{j} U_{n_i} \subset \bigcup_{i=1}^{n} U_i.$$

Hence, for each $x' \in T\left(\bigcup_{i=1}^{n} \bar{U}_i, V_m\right)$, we have $\langle x, x' \rangle \in V_m$ for all $x \in A$. That

shows that $T\left(\bigcup_{i=1}^{n} \bar{U}_i, V_m\right) \subset T(A, V)$ and, hence, $\mathscr{U}'$ is a fundamental system of k-neighborhoods of $0'$ in G'. Since $\mathscr{U}'$ is countable, (a) follows.

(b) G'^k is separable: Let $\{U_n\}$ be as in (a) and let $\{V_n\}$ denote a countable base of open sets in T. (Observe that T, being the quotient group of R by Z, satisfies the second axiom of countability (Theorem 11, §24, Chapter III), because R does.) Let

$$T_n = \{x' \in G' : \langle \bar{U}_n, x' \rangle \subset V_i, i = 1, \ldots, n\}.$$

Then $\{T_n\}$ is a countable family of k-open sets in G'^k. Let M' denote the set of points $\{x_n'\}$, where each x_n' is chosen from T_n if T_n is nonempty. Then M' is clearly countable. We show that M' is dense in G'^k.

For this, let $x' \in G'$ be arbitrary and let $T(F, V)$ be a member of the base of k-neighborhoods of $0'$ in G', in which F is a compact subset of G and V is an open neighborhood of 0 in T. For each $x \in F$, there is a $V_x \in \{V_n\}$, containing $\langle x, x' \rangle$ and such that

(2.1) $$V_x \subset \langle x, x' \rangle + V.$$

Since x' is continuous, there exists an open neighborhood U_x of x such that $U_x \in \{U_n\}$ and

(2.2) $$\langle \bar{U}_x, x' \rangle \subset V_x.$$

As x runs over F, $\{U_x\}$ forms an open covering of F. Compactness of F implies that there is a finite system $\{U_{x_i}\}(1 \leq i \leq m)$ that covers F. Let V_{x_i} be the corresponding V's and let

$$T_m = \{y' \in G' : \langle \bar{U}_{x_i}, y' \rangle \subset V_{x_i}, i = 1, \ldots, m\}.$$

Since $x' \in T_m$, $T_m \neq \varnothing$. Let $y' \in T_m \cap M'$. We show that $y' - x' \in T(F, V)$.

For each $x \in F$, there exists a positive integer n such that $x \in U_{x_n}$ $(1 \leq n \leq m)$. Hence, from (2.2), $\langle x, x' \rangle \in V_{x_n}$. Since $y' \in T_m$, we see that $\langle x, y' \rangle \in V_{x_n}$. Hence,

$$\langle x, y' - x' \rangle = \langle x, y' \rangle - \langle x, x' \rangle \in V_{x_n} - \langle x, x' \rangle \subset V$$

owing to (2.1). This shows that $y' - x' \in T(F, V)$ and proves that M' is k-dense in G'.

46. DUAL GROUPS OF COMPACT AND DISCRETE GROUPS

Theorem 3. *Let G be a Hausdorff locally compact abelian topological group and G'^k its dual group. (a) If G is compact then G'^k is discrete. (b) If G is discrete then G'^k is compact.*

PROOF. For (a) it is sufficient to show that there exists a k-neighborhood of $0'$ (in G') that consists of $0'$ only. Since G is compact, for each open neighborhood V of 0 in T, $T(G, V)$ is a k-neighborhood of $0'$ in G'. Clearly, $T(M, V) \supset T(G, V)$, where M is any compact subset of G. To complete the proof of (a), it is sufficient to show that there exists an open neighborhood V of 0 in T such that $T(G, V) = \{0'\}$. Suppose there exists an $x' \in T(G, V)$ such that $x' \neq 0'$ for a neighborhood V of 0 in T. Let us assume that $V = \left\{ x \in T : -\dfrac{1}{m} < x < \dfrac{1}{m} \right\}$ (cf., Exercise 12(a), Chapter III). Then for some $x \in G$, $\langle x, x' \rangle \neq 0$, and there exists a neighborhood V_1 of 0 in T such that $\sum\limits_{i=1}^{n} V_i \subset V$, where $V_i = V_1 (1 \leq i \leq n)$ and $n \langle x, x' \rangle > \dfrac{1}{m}$. But then we have

$$n \langle x, x' \rangle = \langle nx, x' \rangle \in V$$

or

$$n \langle x, x' \rangle < \frac{1}{m},$$

which is contrary to the choice of n and V_1. Therefore, $x' = 0'$, and (a) is established.

(b) If G is discrete, then $\{0\}$ is a compact open neighborhood of 0 in G. Hence, for each neighborhood V of 0 in T, in view of Proposition 1, §44, $\langle 0, x' \rangle = 0 \in V$ for each $x' \in G'$. That means $T(\{0\}, V) = G'$. But then $Cl_k T(\{0\}, V) = Cl_k G' = G'^k$ is compact (cf., proof of Theorem 1, §45). This proves (b).

Exercises

1. Let R be the additive group of the real numbers endowed with the Euclidean metric topology. Find its character group R'.

2. Find the dual group of R (the additive group of the real numbers) when R is endowed with the discrete topology.

3. Let Z denote the group of integers, endowed with the topology induced from R. Find the dual group of Z.

47. SOME APPLICATIONS OF DUALITY THEORY

In this section we shall discuss some applications of duality theory. It is intended here only to motivate the reader to study other topics to which duality theory is applied. Thus, this section may very well be regarded as an introduction for the study of the books by Rudin,[41] Hewitt and Ross,[19] and Loomis.[29]

Throughout this section G stands for a Hausdorff locally compact abelian (additive) topological group and G' stands for its dual group.

(i) CONVOLUTIONS OF FUNCTIONS

Recall that a Borel measurable complex-valued function f on G (measurable with respect to the Haar measure on G), is said to belong to $L_p(G)$ if

$$\int_G |f(x)|^p \, dx < \infty$$

(integral here is the Haar integral). As usual, we identify functions in $L_p(G)$ that differ from each other on a set of measure zero. In other words, $L_p(G)$ represents the set of equivalence classes of measurable functions f such that the pth power of their absolute value has finite integral.

As in the classical case, $L_p(G)$, $p \geq 1$, are complete normed (or Banach) spaces with the norm:

$$\|f\|_p = \int_G |f(x)|^p \, dx.$$

$L_2(G)$ is a Hilbert space (for definition see §11, Chapter I) with scalar product defined by

$$\langle f, g \rangle = \int_G f(x)\overline{g(x)} \, dx.$$

$L_\infty(G)$ is the space of all bounded measurable functions with the norm:

$$\|f\|_\infty = \sup_{x \in G} |f(x)|,$$

where again two functions are identified if they differ from each other only on a set of measure zero.

The *convolution* function $f * g$ of two complex functions on G is defined as follows:

$$(1) \qquad f * g(x) = \int_G f(x - y)g(y) \, dy,$$

whenever the integral is defined for (almost) all x.

Proposition 4. *The following are elementary facts:*
(a) $f * g = g * f$.
(b) *If $f \in L_1(G)$ and $g \in L_\infty(G)$, then $f * g$ is bounded and uniformly continuous.*
(c) *For $f, g \in L_1(G), f * g \in L_1(G)$ and $\|f * g\|_1 \leq \|f\|_1 \|g\|_1$.*
(d) $(f * g) * h = (f * g) * h$, *for $f, g, h \in L_1(G)$.*

PROOF. (a) Observe that for any Borel measurable subset E, $\mu(-E) = \mu(E)$ because Haar measure is unimodular on an abelian group. Hence, replacing y by $y + x$ in (1), we have

$$f * g(x) = \int_G f(-y)g(y + x)\, dy$$

$$= \int_G f(y)g(-y + x)\, dy = g * f(x).$$

(b) By hypothesis

$$|f * g(x)| \le \|f\|_1 \|g\|_\infty, \qquad \text{for all } x \in G.$$

Hence, $f * g$ is bounded. Furthermore, if $x, z \in G$ then

$$|f * g(x) - f * g(z)| \le \int_G |f(x - y) - f(z - y)||g(y)|\, dy$$

$$\le \|g\|_\infty \int_G |f(x - y) - f(z - y)|\, dy$$

$$\le \|g\|_\infty \int_G |f_{-x}(y) - f_{-z}(y)|\, dy,$$

in which, as usual, f_{-x} is the translate of f, i.e., $f_{-x}(y) = f(y - x)$ for all $y \in G$. (b) will be completed if we show that the mapping: $x \to f_x$, where $f \in L_1$, of G into L_1 is uniformly continuous. By our definition of integral there exists a $g \in C_0(G)$ with compact support K such that $\|f - g\|_1 \le \varepsilon/3$. By uniform continuity of g (Proposition 5, §30, Chapter V) there exists a neighborhood U of 0 in G such that

$$\|g - g_x\|_\infty < \varepsilon/3\mu(K),$$

for all $x \in U$. Hence, $\|g - g_x\|_1 < \varepsilon/3$ and, therefore,

$$\|f - f_x\|_1 \le \|f - g\|_1 + \|g - g_x\|_1 + \|g_x - f_x\|_1 < \varepsilon/3 + \varepsilon/3 + \varepsilon/3 = \varepsilon$$

for all $x \in U$. Hence, if $x - z \in U$,

$$\|f - f_{x-z}\|_1 = \|(f - f_{x-z})_x\|_1 = \|f_x - f_z\|_1 < \varepsilon$$

proves the uniform continuity of the mapping $x \to f_x$ of G into $L_1(G)$. (The same proof holds for the uniform continuity of the mapping $x \to f_x$ of G into $L_p(G)$, $1 \le p \le \infty$.)

(c) Since f is Borel measurable, for each open set V in the complex plane, $f^{-1}(V) \times G$ is a Borel set in $G \times G$. But then by the homeomorphism of the translations $(x, y) \to (x + y, y)$ of $G \times G$ into itself, $E = \{(x, y) : x - y \in f^{-1}(V)\}$ is also a Borel set. Since $(x, y) \in E$ if, and only if, $f(x - y) \in V$, it follows that $f(x - y)$ is a Borel function on $G \times G$. Hence, $f(x - y)g(y)$ is also a Borel function.

By the Fubini theorem (Corollary to Theorem 5, §34, Chapter VI)

$$\int_G \int_G |f(x - y)g(y)|\, dx\, dy \le \|f\|_1 \|g\|_1.$$

Putting $h(x) = \int_G |f(x - y)g(y)| \, dy$, we see that $h \in L_1(G)$. Hence, $h(x) < \infty$ almost everywhere. Therefore, $f * g(x)$ exists almost everywhere. But since $|f * g(x)| \le h(x)$, $f * g \in L_1(G)$ and thus (c) is completed.

(d) In view of (c) we obtain (d) by applying the Fubini theorem:

$$f * (g * h)(x) = \int_G f(x - y)(g * h)(y) \, dy$$

$$= \int_G \int_G f(x - y)g(y - z)h(z) \, dy \, dz$$

$$= \int_G \int_G f(x - y - z)g(y)h(z) \, dy \, dz$$

$$= \int_G f * g(x - z)h(z) \, dz$$

$$= (f * g) * h(x).$$

Corollary 1. $L_1(G)$ is a commutative Banach algebra with convolution as multiplication.

(ii) FOURIER TRANSFORMS OF FUNCTIONS IN $L_1(G)$

As was remarked in Example 4, §19, Chapter III, one can identify the group of the real numbers modulo the subgroup of integers with the group of complex numbers z such that $|z| = 1$. The identity 0 of the former corresponds to the identity 1 of the complex numbers of modulus one. For the sake of convenience, we shall consider the latter in the sequel instead of the former. Thus, a character of G is a continuous homomorphism of G into the group of complex numbers of modulus unity. Since the law of composition in the group T of complex numbers of modulus unity is multiplication, one can rewrite the properties of Proposition 1, §44, Chapter VIII, as follows:

(a) $\langle x + y, x' \rangle = \langle x, x' \rangle \langle y, x' \rangle$;

(b) $\langle x, x'y' \rangle = \langle x, x' \rangle \langle x, y' \rangle$;

(c) $\langle 0, x' \rangle = 1$;

(d) $\langle -x, x' \rangle = \langle x, x'^{-1} \rangle = [\langle x, x' \rangle]^{-1} = \overline{\langle x, x' \rangle}$

($\bar{z}$ denotes the conjugate of the complex number z).

Remark. Sometimes in the literature the "multiplication" in T is actually expressed as addition, i.e.,

$$\langle x, x' + y' \rangle = \langle x, x' \rangle \langle x, y' \rangle,$$

$x \in G$, x', $y' \in G'$ (e.g., see Rudin[41]).

With these remarks, we define the *Fourier transform* of any $f \in L_1(G)$ as the complex-valued function $\hat{f}$ on the dual group G' such that

$$\hat{f}(x') = \int_G f(x)\langle -x, x' \rangle \, dx.$$

Proposition 5. *For any* $x' \in G'$, *the mapping* $\varphi_{x'} : f \to \hat{f}(x')$ *is a nontrivial (i.e., not identically zero) homomorphism of* $L_1(G)$ *into the set of complex numbers. Conversely, every nonzero homomorphism of* $L_1(G)$ *into the set of complex numbers is obtained in this way. Moreover, if* $x' \neq y'$, *then* $\varphi_{x'} \neq \varphi_{y'}$.

PROOF. Let $f, g \in L_1(G)$. Then

$$\widehat{(f * g)}(x') = \int_G f * g(x)\langle -x, x' \rangle \, dx$$

$$= \int_G \int_G f(x - y)g(y)\langle -x, x' \rangle \, dx \, dy$$

$$= \int_G \langle -x, x' \rangle \, dx \int_G f(x - y)g(y) \, dy$$

$$= \int_G g(y)\langle -y, x' \rangle \, dy \int_G f(x - y)\langle -x + y, x' \rangle \, dx$$

$$= \hat{g}(x')\hat{f}(x').$$

Thus the mapping $\varphi_{x'}$ is a multiplicative homomorphism of $L_1(G)$. Since $\varphi_{x'}$ is trivially linear, it is a homomorphism of the Banach algebra $L_1(G)$. Furthermore, since $|\langle -x, x' \rangle| = 1$, $\hat{f}(x') \not\equiv 0$.

For the converse, let φ be a nonzero homomorphism of the group algebra $L_1(G)$ into the algebra of complex numbers. It is easy to see (e.g., see the discussion after Proposition 7, §50, Chapter IX) that φ is a bounded linear functional on $L_1(G)$ such that $\|\varphi\|_\infty = \sup_{\|f\| \leq 1} |\varphi(f)| = 1$. By Theorem 4, §33, Chapter VI, there exists a unique $f_0 \in L_\infty(G)$, $\|f_0\|_\infty = 1$ such that

$$\varphi(f) = \int_G f(x)f_0(x) \, dx$$

for each $f \in L_1(G)$, where the integration is with respect to Haar measure. If $f, g \in L_1(G)$, then

$$\int_G \varphi(f)g(y)f_0(y) \, dy = \varphi(f)\varphi(g)$$

$$= \varphi(f * g) \quad \text{(since } \varphi \text{ is a homomorphism)}$$

$$= \int_G \int_G f(x - y)g(y)f_0(x) \, dy \, dx$$

$$= \int_G g(y) \, dy \int_G f(x - y)f_0(x) \, dx$$

$$= \int_G g(y)\varphi(f_y) \, dy, \quad (f_y(x) = f(x - y)).$$

Hence, $\varphi(f)g(y)f_0(y) = g(y)\varphi(f_y)$ for all y outside a set of measure zero (i.e., almost everywhere). Hence,

$$\varphi(f)f_0(y) = \varphi(f_y),$$

almost everywhere. But since φ is continuous (because φ is bounded) and since f_y is continuous by Proposition 4(b) (see the proof), it follows that $\varphi(f_y)$ is continuous on G for each $f \in L_1(G)$. Since φ is nonzero by assumption, let $f \in L_1(G)$ such that $\varphi(f) \neq 0$. Then $\varphi(f_y)/\varphi(f)$ is a continuous function of y, which coincides with f_0 almost everywhere. Since Haar measure of a nonempty open set is positive, it follows that

$$\varphi(f)f_0(y) = \varphi(f_y)$$

for all $y \in G$, and hence, f_0 is continuous. To show that f_0 is a homomorphism of G into the set of complex numbers, we replace y by $x + y$ in the previous equation, and we obtain

$$\varphi(f)f_0(x + y) = \varphi(f_{x+y}) = \varphi((f_x)_y) = \varphi(f_x)f_0(y)$$
$$= \varphi(f)f_0(x)f_0(y).$$

In other words, $f_0(x + y) = f_0(x)f_0(y)$ for all $x, y \in G$. As for any group homomorphism, $f_0(-x) = [f_0(x)]^{-1}$. Furthermore, since $\|f_0\|_\infty = 1$, it follows that $|f_0(x)| = 1$ for all $x \in G$. This proves that $f_0 \in G'$. By putting $f_0 = x_0' \in G'$, we obtain

$$\varphi(f) = \int_G f(x)f_0(x)\,dx = \int_G f(x)\langle x_0', x\rangle\,dx.$$

To prove the last part of the proposition, if for $x', y' \in G'$, $\varphi_{x'} = \varphi_{y'}$ or $\hat{f}(x') = \hat{f}(y')$ for all $f \in L_1(G)$, then $\langle -x, x'\rangle = \langle -x, y'\rangle$ for almost all $x \in G$. Since x' and y' are continuous, it follows that $\langle -x, x'\rangle = \langle -x, y'\rangle$ for all $x \in G$. In other words, $x' = y'$.

Remarks. 1. The first part of the Proposition says that the Fourier transform of $f * g$ is the product of the Fourier transforms $\hat{f}$ and $\hat{g}$, for $f, g \in L_1(G)$.

2. Fourier transform can be interpreted as convolution. Indeed, if $f \in L_1(G)$ and $x' \in G'$,

$$\hat{f}(x') = \int_G f(x)\langle -x, x'\rangle\,dx$$

$$= \int_G f(x)x'(0 - x)\,dx$$

$$= x' * f = f * x'(0) \qquad \text{(Proposition 4(a))}.$$

3. If $\hat{f}(x) = \overline{f(-x)}$, then $\hat{\bar{f}} = \tilde{f}$.

4. If $\hat{f}(x) = \overline{f(-x)}$, then $\widehat{|f * \tilde{f}|} = |\hat{f} \cdot \tilde{\hat{f}}| = |\hat{f} \cdot \overline{\hat{f}}| = |\hat{f}|^2.$

Proposition 6.   *Let G be a locally compact Hausdorff abelian topological group.*

   (a) *If G is discrete, the algebra $L_1(G)$ has a unit element.*

   (b) *If G is compact, then $L_1(G)$ has a unit element provided G is finite.*

   (c) *If G is not discrete, $L_1(G)$ has approximate unit elements or identities, i.e., there exists a family $\{e_\alpha\}$ of functions in $L_1(G)$ such that for $f \in L_1(G)$ and for an arbitrary $\varepsilon > 0$ there exists $e_\alpha \in \{e_\alpha\}$ such that $\|f * e_\alpha - f\|_1 < \varepsilon$.*

PROOF.   (a) Let G be discrete and let Haar measure on G be normalized, i.e., the measure of each singleton is 1.   Then for $f, g \in L_1(G)$, the convolution can be written as

$$f * g(x) = \sum_{y \in G} f(x - y)g(y).$$

(Here the summation over uncountably many terms is interpreted in the generalized limit sense.   More precisely, let $\{G_\alpha\}$ denote the family of finite subsets G_α of G.   By ordering them by inclusion we obtain a net.   The limit is obtained via this net.)   Now let $e_0 \in L_1(G)$ such that $e_0(0) = 1$ and $e_0(x) = 0$ for $x \neq 0$.   Then, clearly, $f * e_0(x) = f(x)$ and $e_0 * f(x) = f(x)$ for all $x \in G$. Hence, e is the unit element of $L_1(G)$.

   (b) This follows from (a), since a compact discrete space is finite.

   (c) To show that $L_1(G)$ has approximate identities, if G is not discrete, let $f \in L_1(G)$.   From the proof of part (b) of Proposition 4, for each $x \in G$, the mapping $x \to f_x$ is uniformly continuous.   For an arbitrary $\varepsilon > 0$ there exists a neighborhood U of 0 in G such that $\|f - f_x\|_1 < \varepsilon$ for all $x \in U$. Let e_u be a nonnegative Borel function such that $e_u(G \sim U) = 0$ and $\int_G e_u(x) \, dx = 1$.   (It is easy to see that such functions exist.)   Then

$$f * e_u(x) - f(x) = \int_G [f(x - y) - f(x)]e_u(y) \, dy.$$

Hence

$$\|f * e_u - f\|_1 = \int_G |f * e_u(x) - f(x)| \, dx$$

$$\leq \int_G \int_G |f(x - y) - f(x)| \, e_u(y) \, dx \, dy$$

$$\leq \int_U \|f - f_y\|_1 e_u(y) \, dy < \varepsilon.$$

This completes the proof.

   The fact that $L_1(G)$ cannot have a unit element if G is nondiscrete will be established later.

(iii) CONVOLUTION OF MEASURES

   Let μ and γ be any two regular bounded (i.e., $\|\mu\| < \infty$ and $\|\gamma\| < \infty$; cf. Proposition 7(d) below) complex-valued Borel measures on G.   Let $\mu \times \gamma$ denote the product measure on $G^2 = G \times G$.   For each Borel set B in G, the set

$$E_B = \{(x, y) \in G^2 : x + y \in B\}$$

is a Borel set in G^2. For, the mapping $(x, y) \rightarrow (x + y, y)$ of $G \times G$ onto itself is a homeomorphism and, hence, $B \times G$ is mapped onto E_B. Since B is a Borel set, so is E_B.

Define

$$\mu * \gamma(B) = (\mu \times \gamma)(E_B).$$

The set function $\mu * \gamma$ is called the *convolution* of μ and γ.

One can prove the following:

Proposition 7. *Let λ, μ and γ be bounded regular measures on G. Then:*
(a) $\mu * \gamma$ *is a bounded regular measure on G.*
(b) $\mu * \gamma = \gamma * \mu$.
(c) $(\lambda * \mu) * \gamma = \lambda * (\mu * \gamma)$.
(d) $\|\mu * \gamma\| \leq \|\mu\| \|\gamma\|$, *where*
$\|\mu\| = |\mu|(G)$, *and* $|\mu|(B) = \sup \{\sum |\mu(B_i)| \text{ for each finite family } \{B_i\} \text{ of}$
Borel sets such that $\bigcup B_i \subseteq B\}$.

See Rudin[41] for proofs.

The following can be concluded immediately from Proposition 7:

Corollary 2. *The set $M(G)$ of all bounded regular complex-valued measures on G is a commutative Banach algebra with convolution as multiplication. It has a unit element (cf., Proposition 11(a) below).*

Let $\mu \in M(G)$. For each $x' \in G'$ (dual of G),

$$\hat{\mu}(x') = \int_G \langle -x, x' \rangle \, d\mu(x)$$

is called the *Fourier-Stieltjes* transform of μ. Thus, one sees that $\hat{\mu}$ is a complex-valued function on G'. We prove the following:

Proposition 8. (a) *Let $x' \in G'$ be given. Then*
(a) $\hat{\mu}$ *is a bounded and uniformly continuous mapping of G' into the complex numbers.*
(b) *If $\mu, \gamma \in M(G)$,*

$$\widehat{(\mu * \gamma)} = \hat{\mu} \times \hat{\gamma}$$

(c) *The mapping $\mu \rightarrow \hat{\mu}(x')$ is an algebra homomorphism of $M(G)$ into the set of complex numbers.*

PROOF. (a) By the definition of $\hat{\mu}$, we see that

$$|\mu(x')| \leq \|\mu\|$$

for all $x' \in G'$. Hence, $\hat{\mu}$ is bounded. To show that it is uniformly continuous, let $\varepsilon > 0$ be given. By regularity of μ, there exists a compact set C in G such that $|\mu| (G \sim C) < \varepsilon$. For any $x', y' \in G'$, we have

$$|\hat{\mu}(x') - \hat{\mu}(y')| = \left| \int_G \langle -x, x' \rangle \, d\mu(x) - \int_G \langle -x, y' \rangle \, d\mu(x) \right|$$

$$= \left| \int_G \langle -x, x' \rangle \{1 - \langle -x, y' \rangle [\langle -x, x' \rangle]^{-1}\} \, d\mu(x) \right|$$

$$\leq \int_G |1 - \langle x, y' - x' \rangle| \, d|\mu|(x)$$

$$\leq \int_C |1 - \langle x, y' - x' \rangle| \, d|\mu|(x)$$

$$+ \int_{G \sim C} |1 - \langle x, y' - x' \rangle| \, d|\mu|(x).$$

Since $T(C, \varepsilon)$ is a member of the subbase of the topology k, where
$$T(C, \varepsilon) = \{x' \in G' : |\langle x, x' \rangle| < \varepsilon, \text{ for all } x \in C\},$$
we see that if $y' - x' \in T(C, \varepsilon)$, then
$$|1 - \langle x, y' - x' \rangle| = |\langle -x, x' \rangle - \langle -x, y' \rangle| < \varepsilon$$
for $x \in C$. Hence,

$$\int_C |1 - \langle x, y' - x' \rangle| \, d \, |\mu| \, (x) \leq \varepsilon \, |\mu| \, (C) \leq \varepsilon \, \|\mu\|.$$

Furthermore,

$$\int_{G \sim C} |1 - \langle x, y' - x' \rangle| \, d \, |\mu| \, (x) \leq 2 \, |\mu| \, (G \sim C) < 2\varepsilon.$$

Therefore, $|\hat{\mu}(x') - \hat{\mu}(y')| \leq \varepsilon(2 + \|\mu\|)$, whenever $y' - x' \in T(C, \varepsilon)$. This proves the uniform continuity of $\hat{\mu}$.

(b) To prove this, we first observe that if χ_B is a characteristic function of a Borel set B in G, then by the definition of convolution $\mu * \gamma(B)$ we have

(A) $$\int_G \chi_B \, d(\mu * \gamma) = \int_G \int_G \chi_B(x + y) \, d\mu(x) \, d\gamma(y).$$

Actually, this is equivalent to the definition of $\mu * \gamma(B)$.

Since the integral is a linear functional, (A) is true for each simple function (i.e., a finite linear combination of characteristic functions). Hence, since each bounded Borel function is the uniform limit of a sequence of simple functions, (A) is true for each bounded Borel function instead of χ_B. Thus, we have

$$\widehat{(\mu * \gamma)}(x') = \int_G \langle -z, x' \rangle \, d(\mu * \gamma)(z)$$

$$= \int_G \int_G \langle -x - y, x' \rangle \, d\mu(x) \, d\gamma(y)$$

$$= \int_G \langle -x, x' \rangle \, d\mu(x) \int_G \langle -y, x' \rangle \, d\gamma(y)$$

$$= \hat{\mu}(x')\hat{\gamma}(x').$$

This proves (b).

(c) Since $\mu \to \hat{\mu}(x')$ is additive, the result follows from (b).

Proposition 9. $L_1(G)$ *can be regarded as a subalgebra of* $M(G)$.

PROOF. For each $f \in L_1(G)$, let

$$\mu_f(E) = \int_E f(x)\, dx.$$

Then μ_f is absolutely continuous with respect to Haar measure on G (cf., Halmos[16]). Conversely, by the Radon-Nikodym theorem (Theorem 3, §33, Chapter VI) every absolutely continuous $\mu \in M(G)$ is u_f for some $f \in L_1(G)$.

Since $L_1(G)$ is the set of equivalence classes of functions f differing from each other at most on sets of measure zero, and since their integrals with respect to Haar measure are finite, we conclude that there is a one-to-one correspondence between f and μ_f. This proves that $L_1(G)$ is a subset of $M(G)$.

Furthermore, we see that, for all $x' \in G'$, $\hat{f}(x') = \mu_f(x')$ and $\|f\|_1 = \|\mu_f\|$. This shows that $L_1(G)$ is a subalgebra of $M(G)$, and the L_1-norm on $L_1(G)$ coincides with the norm induced from $M(G)$.

Remark. In view of Proposition 9, whenever $\mu \in L_1(G)$ or $\mu = \mu_f$ we see that $f * \mu \in L_1(G)$ for any $f \in L_1(G)$, as is the case for $f * g$ where $f, g \in L_1(G)$ (Proposition 4(c)).

Since G'^k is locally compact, we may speak of Haar measure on G'^k. As for G, one can define $M(G')$. We have the following *uniqueness theorem:*

Proposition 10. *If* $\mu \in M(G')$ *and if, for each* $x \in G$,

$$\int_{G'} \langle x, x' \rangle \, d\mu(x') = 0$$

then $\mu \equiv 0$. *Similarly, if* $\mu \in M(G)$ *and for each* $x' \in G'$,

$$\int_G \langle x, x' \rangle \, d\mu(x) = 0$$

then $\mu \equiv 0$.

PROOF. We observe that the set $T(G')$ of all Fourier transforms $\hat{f}$, when f runs over $L_1(G)$, is dense in $C_\infty(G')$. (See the proof of part (c) of Proposition 6.) Now since, for every $f \in L_1(G)$,

$$\int_{G'} \hat{f}(x') \, d\mu(x') = \int_{G'} \int_G f(x) \langle -x, x' \rangle \, dx \, d\mu(x')$$

$$= \int_G f(x) \, dx \int_{G'} \langle -x, x' \rangle \, d\mu(x') = 0$$

by hypothesis; and since $T(G')$ is dense in $C_\infty(G')$, it follows that

$$\int_{G'} g(x') \, d\mu(x') = 0$$

for each $g \in C_\infty(G')$. Hence $\mu \equiv 0$.

The other part follows by symmetry between G and G'.

Corollary. *$M(G)$ and $L_1(G)$ are semisimple (cf., §50, Chapter IX, for the definition) commutative Banach algebras.*

PROOF. We have already shown (Corollaries 2 and 1) that $M(G)$ and $L_1(G)$ are commutative Banach algebras. To show first that $M(G)$ is semisimple, let $\mu \in M(G)$. For each $x' \in G'$, the mapping $\mu \rightarrow \hat{\mu}(x')$ is a homomorphism of the Banach algebra $M(G)$ into the algebra of complex numbers (Proposition 8(c)). Hence by Proposition 10, $M(G)$ is semisimple. Furthermore, $L_1(G)$ can be identified with a subalgebra of $M(G)$ (Proposition 9). Hence Proposition 10 holds for elements in $L_1(G)$ and, thus, follows the semisimplicity of $L_1(G)$.

Proposition 11. (a) *$M(G)$ is a commutative semisimple Banach algebra with a unit element.*
(b) *If G is a nondiscrete locally compact abelian topological group, then $L_1(G)$ does not have a unit element.*

PROOF. (a) We have to show only that $M(G)$ has a unit element. Let μ_0 denote the unit measure concentrated at the identity of G, i.e., $\mu_0(E) = 1$ or 0 according as $0 \in E$ or $0 \notin E$. Then for any $\mu \in M(G)$, $\mu * \mu_0 = \mu$ and (a) is established.

(b) If G is nondiscrete, G' is noncompact (Theorem 3, §46). Let $T(G')$ denote the set of all Fourier transforms of elements in $L_1(G)$. As follows from the discussion at the end of Chapter IX and from Proposition 5 above, the Fourier transform $\hat{f}$ of $f \in L_1(G)$ is precisely the Gelfand mapping and $T(G')$ is a subalgebra of $C_\infty(G')$, the set of all continuous complex-valued mappings of G' that vanish at infinity. (Observe that the maximal ideal space (cf., §50, Chapter IX) of $L_1(G)$ is G'.) Since $L_1(G)$ is semisimple, by Propositions 5 and 8, Chapter IX, $T(G')$ is isomorphic algebraically with $L_1(G)$. Since $T(G') \subset C_\infty(G')$ and $T(G')$ contains no nonzero constant functions, $T(G')$ does not have a unit element. Hence, $L_1(G)$ does not have a unit element.

(iv) POSITIVE-DEFINITE FUNCTIONS
A function p on G is said to be *positive-definite* if

$$(1) \qquad \sum_{i,j=1}^{n} c_i \bar{c}_j p(x_i - x_j) \geq 0$$

for every choice of $\{x_i \in G\}(1 \leq i \leq n)$, and for every choice of complex numbers $\{c_i\}(1 \leq i \leq n)$, ($\bar{c}$ is the complex conjugate of c).

Proposition 12. *For every positive-definite function p on G,*

(a) $p(-x) = \overline{p(x)}$;
(b) $|p(x)| \leq p(0)$;
(c) $|P(x) - p(y)|^2 \leq 2p(0)\mathscr{R}\{p(0) - p(x - y)\}$; *in which $\mathscr{R}\{z\}$ is the real part of the complex number z;*

(d) *Every p is bounded and uniformly continuous if p is continuous at $0 \in G$.*

PROOF. (a) Let $n = 2$. Then (1) gives

$$(2) \qquad (|c_1|^2 + |c_2|^2)p(0) + c_1 \bar{c}_2 p(x_1 - x_2) + c_2 \bar{c}_1 p(x_2 - x_1) \geq 0.$$

Putting $c_1 = c_2 = 1$, $x_2 = 0$, and $x_1 = x$ in (2) we obtain

$$2p(0) + p(x) + p(-x) \geq 0.$$

If $\alpha = p(x) + p(-x)$, then α must be real. Moreover, if $c_2 = \sqrt{-1} = i$, $c_1 = 1$, $x_1 = 0$, and $x_2 = x$ then from (2), $i[p(x) - p(-x)]$ must be real, or $p(x) - p(-x)$ must be purely imaginary. Let $p(x) - p(-x) = i\beta$, where β is real. Then we see that $p(x) = \frac{1}{2}(\alpha + i\beta) = \overline{p(-x)}$.

(b) In (2), put $c_1 = 1$, $c_2 = -|p(x)|/p(x)$, where $x_2 = x$ and $x_1 = 0$. Then (b) follows.

(c) Let $n = 3$, $x_1 = 0$, $x_2 = x$, $x_3 = y$, $c_1 = 1$, $c_3 = -c_2$, and $c_2 = \lambda |p(x) - p(y)|/[p(x) - p(y)]$, in which λ is real. Then from (1) we obtain

$$(3) \qquad p(0) + 2|p(x) - p(y)|\,\lambda + 2[p(0) - \mathscr{R}p(x - y)]\lambda^2 \geq 0.$$

Hence, the discriminant of the quadratic polynomial (3) in λ must be non-negative and thus we obtain (c).

(d) This follows from (b) and (c).

Proposition 13. (i) *If $f \in L_2(G)$, then $f * \tilde{f}$ is positive-definite, where $\tilde{f}(x) = \overline{f(-x)}$.*

(ii) *Every character x' in G' is a positive-definite function.*

(iii) *For each $v \in M(G')$, $v \geq 0$, $p(x) = \displaystyle\int_{G'} \langle x, x' \rangle \, dv(x')$ is positive-definite and continuous.*

PROOF. (i)

$$\sum c_i \bar{c}_j f * \tilde{f}(x_i - x_j) = \sum c_i \bar{c}_j \int_G f(x_i - x_j - x)\overline{f(-x)}\, dx$$

$$= \sum c_i \bar{c}_j \int_G f(x_i - x)\overline{f(x_j - x)}\, dx$$

$$= \int_G |\sum c_i f(x_i - x)|^2 \, dx \geq 0.$$

(ii) The proof is simple.
(iii) We see that

$$\sum c_i \bar{c}_j p(x_i - x_j) = \int_{G'} \sum c_i \bar{c}_j \langle x_i - x_j, x' \rangle \, dv(x')$$

$$= \int_{G'} \left| \sum c_i \langle x_i, x' \rangle \right|^2 dv(x') \geq 0.$$

Hence, p is positive-definite. The continuity of p follows in the same way as that of $\hat{\mu}$ in Proposition 8(a).

One of the key theorems in this area is the following:

Theorem 4 (*Bochner-Weil*). *A continuous function p on G is positive-definite if, and only if, there is a nonnegative measure $\nu \in M(G')$ such that*

$$p(x) = \int_{G'} \langle x, x' \rangle \, d\nu(x')$$

for all $x \in G$.

PROOF. The "only if" part has already been established (Proposition 13(iii)).

For the "if" part, let p be positive-definite. By Proposition 12, p is a bounded function. Without loss of generality we may assume $p(0) = 1$. Hence, $\|p\|_\infty = 1$. For each $f \in C_0(G)$, first we show that

$$(1) \qquad \int_G \int_G f(x)\overline{f(y)}p(x - y) \, dx \, dy$$

is nonnegative. Let M be the compact support of f. It is clear that the integrand in (1) is uniformly continuous on $M \times M$. Let $\{E_i\}(1 \le i \le n)$ be a finite family of pairwise disjoint subsets of G such that $M = \bigcup_{i=1}^{n} E_i$ and such that

$$(2) \qquad \sum_{i,j} f(x_i)\overline{f(x_j)}p(x_i - x_j)\mu(E_i)\mu(E_j) \qquad (\mu \text{ is Haar measure})$$

is as close to (1) as we wish. Since p is positive-definite, the sum in (2) is nonnegative. Hence, the integral in (1) is nonnegative for each $f \in C_0(G)$. Since $C_0(G)$ is dense in $L_1(G)$, (1) is nonnegative for each $f \in L_1(G)$. Now we define a linear functional on $L_1(G)$ as follows:

$$(3) \qquad T_p(f) = \int_G f(x)p(x) \, dx$$

for $f \in L_1(G)$. For $f, g \in L_1(G)$, define a functional on $L_1(G) \times L_1(G)$ as follows:

$$(4) \qquad [f, g] = T_p(f * \tilde{g}),$$

in which $\tilde{g}(x) = \overline{g(-x)}$, $x \in G$. (Observe that $\tilde{g} \in L_1(G)$ as well as $f * \tilde{g} \in L_1(G)$.) It is easy to check that $[f, g]$ is linear in f, $[f, g] = \overline{[g, f]}$, and $[f, f] \ge 0$. This shows that $[f, g]$ is a scalar product function. Hence by Schwarz's inequality,

$$(5) \qquad |[f, g]|^2 \le [f, f][g, g].$$

Now choose a symmetric open neighborhood U of the identity in G. Put $g \equiv [\mu(U)]^{-1}\chi_u$, in which $\mu(U)$ is the Haar measure of U and χ_u the characteristic function of U. Then from (3) and (4) we obtain

$$(6) \qquad [f, g] - T_p(f) = \int_G f * \tilde{g}(x)p(x)\, dx - \int_G f(x)p(x)\, dx$$

$$= \int_G \int_G f(x - y)\overline{g(-y)}p(x)\, dx\, dy - \int_G f(x)p(x)\, dx$$

$$= \int_G \int_U f(x - y)[\mu(U)]^{-1}p(x)\, dx\, dy - \int_G f(x)p(x)\, dx$$

$$= \int_G [\mu(U)]^{-1} \int_U f(x)p(x - y)\, dx\, dy - \int_G f(x)p(x)\, dx$$

$$= [\mu(U)]^{-1} \int_G f(x) \int_U [p(x - y) - p(x)]\, dy\, dx.$$

Similarly, from (4) we have

$$(7) \qquad [g, g] - 1 = [\mu(U)]^{-2} \int_U \int_U [p(x - y) - 1]\, dx\, dy.$$

Since p is continuous by hypothesis and, hence, uniformly continuous (Proposition 12(d)), (6) and (7) can be made as small as we please by taking U small enough. Thus, from (5) we have

$$|T_p(f)|^2 \le [f, f] = T_p(f * \tilde{f})$$

for each $f \in L_1(G)$. If we put $h = f * \tilde{f}$ and $h^n = h^{n-1} * h$ for $n \ge 2$, from the last inequality we obtain

$$(8) \qquad |T_p(f)|^2 \le T_p(h) \le [T_p(h^2)]^{1/2} \le \cdots \le [T_p(h^{2^n})]^{2^{-n}} \le \|h^{2^n}\|_1^{2^{-n}},$$

since $\|T_p\| = 1$ because $\|p\|_\infty = 1$. By the formula following Theorem 1, §49, Chapter IX,

$$\lim_{n \to \infty} \|h^{2^n}\|_1^{2^{-n}} = r(h),$$

where $r(h)$ is the spectral radius of h. By Proposition 8(b), §50, Chapter IX, $r(h) = \|\hat{h}\|_\infty$. Hence, in view of Remark 4 following Proposition 5, from (8) we obtain

$$|T_p(f)|^2 \le \|\hat{h}\|_\infty = \|\hat{f}\|_\infty^2,$$

or

$$|T_p(f)| \le \|\hat{f}\|_\infty.$$

This shows that T_p can be regarded as a bounded linear functional on $T(G') = \{\hat{f} : f \in L_1(G)\}$ with sup norm (Remark (b) following Proposition 8, Chapter IX).

Since $T(G')$ is dense in $C_\infty(G')$, we can extend T_p on the whole of $C_\infty(G')$, preserving its norm. But then by Theorem 4, §33, Chapter VI, there exists a

$\nu \in M(G')$, $\|\nu\| = 1$ such that

(9)
$$T_p(f) = \int_{G'} \hat{f}(-x')\, d\nu(x')$$
$$= \int_{G'} \int_G f(x)\langle -x, -x'\rangle\, dx\, d\nu(x')$$
$$= \int_G f(x)\, dx \int_{G'} \langle x, x'\rangle\, d\nu(x').$$

From (3) and (9) we have

$$\int_G f(x)p(x)\, dx = \int_G f(x)\, dx \int_{G'} \langle x, x'\rangle\, d\nu(x').$$

Hence,

(10)
$$p(x) = \int_{G'} \langle x, x'\rangle\, d\nu(x')$$

almost everywhere. But both functions $p(x)$ and $\int_{G'} \langle x, x'\rangle\, d\nu(x')$ are continuous. Hence, (10) holds everywhere.

To complete the proof, we must show that $\nu \geq 0$. Putting $x = 0$ in (10), we obtain

$$1 = p(0) = \int_{G'} \langle 0, x'\rangle\, d\nu(x') = \int_{G'} d\nu(x')$$
$$= \nu(G') \leq \|\nu\| = 1.$$

Hence, $\nu(G') = \|\nu\| = 1$ implies $\nu \geq 0$.

Remark. The representation of p given in (10) is unique as follows from the uniqueness theorem (Proposition 10).

The characterization of positive-definite functions given by the Bochner-Weil theorem plays an important role in harmonic analysis. For instance, the proofs of two important theorems (the Plancherel and Inversion theorems), which are treated in the next section, depend on the Bochner-Weil theorem.

(v) THE PLANCHEREL THEOREM

One of the most interesting theorems in harmonic analysis is the Plancherel theorem. We shall prove this theorem via the so-called inversion theorem.

Theorem 5 (*Inversion theorem*). *Let $f \in L_1(G)$ be such that for some $\mu \in M(G')$ and each $x \in G$,*

(1)
$$f(x) = \int_{G'} \langle x, x'\rangle\, d\mu(x').$$

Then the Fourier transform $\hat{f}$ of f is in $L_1(G')$.

Moreover, for a fixed Haar measure on G, the Haar measure on G' can be so normalized that the inversion formula

(2)
$$f(x) = \int_{G'} \hat{f}(x')\langle x, x'\rangle\, dx'$$

holds for every $f \in L_1(G)$ satisfying (1).

PROOF. Let the measure μ with respect to which $f(x)$ is represented by the integral (1), be denoted by μ_f for the sake of clarity. Then, if $f \in L_1(G)$ satisfies (1) and if $h \in L_1(G)$, we have

$$(3) \qquad h * f(0) = \int_G h(-x) f(x)\, dx = \int_{G'} \hat{h}(x')\, d\mu_f(x').$$

If g is another function in $L_1(G)$ and if $g(x)$ satisfies (1) then from (3) we have

$$\int_{G'} \hat{h}(x') \hat{g}(x')\, d\mu_f(x') = ((h * g) * f)(0) = (h * f) * g(0) = \int_{G'} \hat{h}(x') \hat{f}(x')\, d\mu_g.$$

Since the set $T(G')$ of all Fourier transforms f, where f runs over $L_1(G)$, is dense in $C_\infty(G')$, as remarked in the proof of Proposition 10, it follows that

$$(4) \qquad \hat{g}\, d\mu_f = \hat{f}\, d\mu_g$$

for $f, g \in L_1(G)$ having the integral representation in (1).

Now to define a linear functional on $C_0(G')$, let $\varphi \in C_0(G')$, with compact support K. For each $x_0' \in K$ there exists a function $\varphi \in C_0(G)$ such that $\hat{\varphi}(x_0') \neq 0$, because $C_0(G)$ is dense in $L_1(G)$ (Proposition 5). Hence, $\widehat{(\varphi * \tilde{\varphi})}(x_0')$ > 0 and $\widehat{(\varphi * \tilde{\varphi})} \geq 0$. Owing to the compactness of K there is a finite family $\{\varphi_i\}(1 \leq i \leq n)$ of functions in $C_0(G)$ such that $g = \sum_{i=1}^{n} \varphi_i * \tilde{\varphi}_i$ has $\hat{g} > 0$ on K. Clearly, $g \in C_0(G)$. Since $C_0(G) \subset L_2(G)$, by Proposition 13(i), g is also positive-definite. Hence, g has the integral representation (1) (Theorem 4). Let μ_g be the measure associated by the Bochner-Weil theorem to g in connection with this representation.

For each function $\psi \in C_0(G')$, define

$$T(\psi) = \int_{G'} (\psi/\hat{g})\, d\mu_g.$$

Then T is well-defined. Moreover, if g is replaced by any other function f such that $\hat{f}(K) \neq 0$, then $T(\psi)$ remains unchanged because by (4),

$$T(\psi) = \int \frac{\psi}{\hat{f}\hat{g}} \hat{f}\, d\mu_g = \int \frac{\psi}{\hat{f}\hat{g}} \hat{g}\, d\mu_f.$$

T is clearly a linear functional. Since g is positive-definite, $\mu_g \geq 0$ and, hence, $T(\psi) \geq 0$ for $\psi \geq 0$. Since there are ψ and μ_f such that $\int \psi\, d\mu_f \neq 0$, for $g = \sum_{i=1}^{n} \varphi_i * \tilde{\varphi}_i$ as chosen above, we have

$$(5) \qquad T(\psi \hat{f}) = \int_{G'} (\psi \hat{f}/\hat{g})\, d\mu_g = \int_{G'} (\psi/\hat{g})\hat{g}\, d\mu_f = \int_{G'} \psi\, d\mu_f \neq 0$$

by using (4). Hence, T is not identically zero.

Fixing $\psi \in C_0(G')$ and $x'_0 \in G'$ and constructing g as above so that $\hat{g} > 0$ on the support K of ψ as well as on $K + x'_0$, we have

$$\{\overline{\langle -x', x'_0 \rangle g(x')}\} = \hat{g}(x' + x'_0)$$

and for any Borel set B in G'.

$$\mu_f(B) = \mu_g(B - x'_0).$$

Thus,

$$T(\psi_{x'_0}) = \int_{G'} [\psi(x' - x'_0)/\hat{g}(x)] \, d\mu_g(x')$$
$$= \int_{G'} [\psi(x')/\hat{f}(x')] \, d\mu_f(x') = T(\psi).$$

This means that T is a left-invariant linear transformation on $C_0(G')$. Hence, by the theorem on Haar measure, T is a Haar integral, i.e.,

$$(6) \qquad\qquad T(\psi) = \int_{G'} \psi(x') \, dx',$$

in which the integral is with respect to the Haar measure on G', and $x' \in G'$.

Now if f has the representation (1) and $\psi \in C_0(G')$, by (5) and (6) we have

$$\int_{G'} \psi \, d\mu_f = T(\psi \hat{f}) = \int_{G'} \psi \hat{f} \, dx',$$

which holds for every $\psi \in C_0(G')$. Hence,

$$(7) \qquad\qquad \hat{f} \, dx' = d\mu_f,$$

for $f \in L_1(G)$ having the representation (1). Since μ_f is a finite measure, it follows that $\hat{f} \in L_1(G')$. Thus, by substituting (7) into (1) we have

$$f(x) = \int_{G'} \hat{f}(x') \langle x, x' \rangle \, dx',$$

which is (2). This completes the proof.

From this theorem we derive the following:

Theorem 6 (*Plancherel*).   *If $f \in L_1(G) \cap L_2(G)$, then $\hat{f} \in L_2(G')$. Furthermore,*

$$\|f\|_2^2 = \int_G |f(x)|^2 \, dx = \int_{G'} |\hat{f}(x')| \, dx' = \|\hat{f}\|_2^2.$$

In other words, the Fourier transform, regarded as a mapping, is an isometry of $L_1(G) \cap L_2(G)$ onto a subset of $L_2(G')$.

PROOF.   Let $f \in L_1(G) \cap L_2(G)$. Then $g = f * \tilde{f} \in L_1(G)$ (Proposition 4(c)). Moreover, g is positive-definite (Proposition 13(i)), and also continuous (Proposition 4(b)). Further, by Remark 4 following Proposition 5, we have

$$|\hat{g}(x')| = |\hat{f}(x')|^2.$$

Therefore, by the inversion theorem, it follows that

$$\int_G |f(x)|^2 \, dx = \int_G f(x)\overline{f(x)} \, dx$$

$$= \int_G f(x)\tilde{f}(-x) \, dx$$

$$= f * \tilde{f}(0) = \int_{G'} \hat{g}(x')\langle 0, x' \rangle \, dx'$$

$$= \int_{G'} \hat{g}(x') \, dx' = \int_{G'} |\hat{f}(x')|^2 \, dx'.$$

In other words, $\|f\|_2^2 = \|\hat{f}\|_2^2$.

Remark. It can be shown that the image, under the mapping $f \to \hat{f}$, of $L_1(G) \cap L_2(G)$ is dense in $L_2(G')$. Thus, this mapping can be extended uniquely on the whole of $L_2(G)$ onto $L_2(G')$. The fact that this extension of the isometry $f \to \hat{f}$ onto the whole of $L_2(G)$ exists is sometimes called the *Plancherel transform*. In this sense one can conclude that each function in $L_2(G')$ is the Plancherel transform of some function in $L_2(G)$.

A special form of the Plancherel theorem is called the *Parséval equality*, which is derived as follows:

Let $G = T$ (Torus group, i.e., the set of all complex numbers z such that $|z| = 1$). Then we see that G' is the group of integers. The Haar measure on G' is the point-measure and, hence, integration is just the summation of series. If we normalize the Haar measure on G by dividing by 2π, we obtain from Plancherel's theorem,

$$\sum_{n=1}^{\infty} |\hat{f}_n|^2 = \frac{1}{2\pi} \int_0^{2\pi} |f(x)|^2 \, dx.$$

This is known as the Parséval equality.

It turns out that Plancherel's theorem is very deep. Attempts have been made to prove the analog of this theorem for nonabelian locally compact groups. For further information on this topic, the reader is referred to Mackey's address (Mackey[32]).

Apart from those applications of duality theory that are discussed in this section, one can discuss the dual of the dual groups of G. If G is locally compact, we have seen that G'^k is also locally compact. Thus, it follows that the dual $G'^{k'k}$ of G'^k is also locally compact. The Pontrjagin duality theorem asserts that $G'^{k'k} = G$. We shall not go into details of this theorem. The reader can consult any one of the following books: Pontrjagin,[37] Rudin,[41] Hewitt and Ross.[19] (See Exercise 3 below.)

Since the dual group G'^k of G is locally compact and since G'^d endowed with the discrete topology is a discrete Hausdorff locally compact abelian topological group, it follows that the dual group $G'^{d'k}$ of G'^d is compact. If we identify G as a subgroup of $G'^{d'k}$ (cf, Exercise 3 below), we can conclude that the closure $\bar{G}$ of G taken in $G'^{d'k}$ is a compact topological group in which G is dense. $\bar{G}$ is called the *Bohr compactification* of G. The Bohr compactification plays an important role in almost periodic functions. The study of this topic is beyond the scope of this book. The above-mentioned books can be consulted for further references.

Exercises

1. Let H be a closed subgroup of a locally compact Hausdorff abelian topological group G and let G' be the dual group of G. Let $H' = \{x' \in G' : \langle x, x' \rangle = 1, \text{ for all } x \in H\}$. Show that H' is a closed subgroup of G'^k; and H' and G'/H' are dual groups of G/H and H, respectively.

2. (a) Find the dual group of a finite product of locally compact abelian Hausdorff groups.

(b) Find the dual groups of R^n, Z^n and T^n (cf., Examples 5, 6, §19).

3. Let G be a locally compact Hausdorff abelian topological group and G' its dual group. Let τ be a mapping: $x \to x''$ of G into $G'^{k'}$ defined by $\tau(x) = x''(x') = \langle x, x' \rangle$ for each $x \in G$ while x' runs over G'. Show that:

(a) τ is a homeomorphism and isomorphism of G onto $G'^{k'k}$. (Pontrjagin duality theorem.)

(b) τ is a continuous homomorphism of G onto a dense subgroup of the Bohr compactification.

4. Let G be a topological group as in Exercise 3, and let U be a neighborhood of the identity in G. Then U contains a closed subgroup H of G such that G/H is homeomorphic and isomorphic with $T^n \times F$, for some finite integer $n \geq 0$ and for some finite abelian group F. (Use this to prove Exercise 5, §29, Chapter IV.)

5. Let G be a compact Hausdorff abelian group.

(a) Show that G' is an orthonormal family of functions in $C(G)$.

(b) Show that the algebra of finite linear combinations of elements in G' is dense in the Banach algebra $C(G)$. (Hint: Use Theorem 7, §11, Chapter I.)

Introduction
to
Banach Algebras

An elegant way of proving most of the theorems of §47, is via Banach algebras. The topic of Banach algebras has been of great interest to mathematicians for the last few decades. One reason for this is that in Banach algebras, ideas from algebra, topology and analysis converge; another is that the tools of Banach algebras are sharp enough to interpret some outstanding theorems of classical analysis. The purpose of this chapter is to demonstrate quickly, without going into details, how some of the theorems and notions discussed in §47 can be interpreted in terms of quantities associated with Banach algebras. A detailed treatment of this topic can be found in Loomis,[29] Hille and Phillips,[49] and Rickart[50] Also the address of Gelbaum[13] is very relevant for this purpose.

48. DEFINITIONS AND EXAMPLES OF
BANACH ALGEBRAS

An algebra A (§11, Chapter I) that is a normed (or Banach) space is called a normed (or Banach) algebra if for $x, y \in A$,

$$\|xy\| \leq \|x\|\,\|y\|.$$

This inequality implies that the product in A is a continuous mapping of two variables together.

In view of this definition, one can define a topological algebra more generally as follows: An algebra A is said to be a *topological algebra* if A

is a topological linear space (Exercise C, §18, Chapter II) in which the product mapping: $(x, y) \rightarrow xy$ of $A \times A$ into A is continuous in both variables together. (We shall not make use of this definition in the sequel.)

A normed algebra is said to be real or complex according as the field of scalars is real or complex, as is the case with a normed space. We shall be concerned mainly with complex Banach algebras in the sequel. Most of our considerations have an immediate interpretation for the real case, too.

A Banach algebra need not have a multiplicative identity. That is the reason that one is required to have a theory of Banach algebras without identity if one wishes to include some well-known Banach algebras with this defect. For example, the group algebra $L_1(G)$ discussed in §47 does not always have an identity. Whenever there is an identity, the theory is more easily derived.

A topological algebra need not be commutative. For example, the normed algebra $\mathbf{M}(C)$ of $n \times n$ matrices is not commutative. Not as much is known about noncommutative algebras as is known about commutative algebras. We shall discuss only commutative normed algebras in the sequel.

Normed algebras can be classified as follows: (i) *function algebras*, (ii) *operator algebras*, and (iii) *group algebras*.

We give examples of each class.

I. Let T be a topological space and let $C(T)$ denote the set of all continuous bounded complex-valued functions on T. Let addition and multiplication in $C(T)$ be point-wise, i.e., $f + g(x) = f(x) + g(x)$ and $fg(x) = f(x)g(x)$. Then $C(T)$ is an algebra. Define

$$\|f\| = \sup_{x \in T} |f(x)|$$

for each $f \in C(T)$. With this norm $C(T)$ is a commutative Banach algebra with an identity—the identity being the complex function mapping T onto 1.

The algebra $C(T)$ is an example of a function algebra. For different choices of T, one finds different algebras. For example, T can be taken to be a single point, a closed bounded interval, the unit disc $= \{z : z$ complex, $|z| \leq 1\}$, or more generally a compact Hausdorff space. If T is a single element, $C(T)$ coincides with the set of complex numbers.

II. Let E be a Banach space. The set $B(E)$ of all bounded operators of E into itself is an algebra with composition as multiplication. For each $T \in B(E)$, let

$$\|T\| = \sup_{\|x\| \leq 1} \|T(x)\|.$$

With this norm, $B(E)$ is a Banach algebra.

The algebra $B(E)$ is an example of an *operator algebra*. For different choices of E, one finds different algebras. For example, if $E = C^n$, n-dimensional complex Euclidean space (§27, Chapter IV), $B(E)$ is nothing but the algebra of $\mathbf{M}_n(C)$ of $n \times n$ matrices, with complex coefficients. Furthermore, if H is a Hilbert space (§11, Chapter I), $B(H)$ is a special Banach algebra that plays an important role in spectral theory. Moreover, in $B(H)$ there is an additional structure available, viz., for each $T \in B(T)$ there is a conjugate T^* of T defined by the following formula. If $\langle x, y \rangle$ denotes the scalar product in H, then for $T \in B(H)$,

$$\langle T(x), y \rangle = \langle x, T^*(y) \rangle,$$

$x, y \in H$.

A subalgebra A of $B(H)$ is said to be *self-adjoint* if for each $T \in A$, $T^* \in A$. A self-adjoint closed subalgebra or in other words, a self-adjoint Banach subalgebra of the Banach algebra $B(H)$ is said to be a *C*-algebra*.

III. For our purpose in this book, the most important class of algebras is the class of group algebras.

Let G be a locally compact Hausdorff topological abelian group. In view of Corollary 1, §47, $L_1(G)$ is a Banach algebra with convolution as multiplication, viz., for $f, g \in L_1(G)$,

$$f * g(x) = \int_G f(x - y)g(y)\,dy$$

(see §47(i)), in which integration is performed with respect to Haar measure. The fact that $f * g \in L_1(G)$ has been established in Proposition 4 (§47). The fact that $L_1(G)$ is a Banach space is well-known. The algebra $L_1(G)$ is an example of a group algebra. It does not have an identity but it does have approximate identities (Proposition 6, §47, Chapter VIII). If G is discrete then $L_1(G)$ has an identity.

If G is a finite group, Haar measure is a point measure (i.e., it assigns constant, nonzero measure to each element of G) (see Halmos[16]) and thus the convolution is (if each element is assigned measure 1)

$$f * g(x_i) = \sum_{j=1}^{n} f(x_i - x_j)g(x_j),$$

in which n is the number of elements x_i in G.

If $G = Z$, the group of integers, Haar measure is the point measure and the convolution is

$$f * g(n) = \sum_{m=-\infty}^{\infty} f(n - m)g(m).$$

If $G = R$, the additive group of the real numbers, Haar measure is Lebesgue measure and the convolution has the well-known form:

$$f * g(x) = \int_{-\infty}^{\infty} f(x - y)g(y)\,dy.$$

49. THE GELFAND-MAZUR THEOREM

To establish the Gelfand-Mazur theorem, we need some results concerning the regularity and spectrum, etc., of an element in a normed algebra.

For this we consider the following notions:

Let A be a commutative normed algebra. Suppose A has a unit element. An element $x \in A$ is said to be *regular* if there exists $y \in A$ such that $xy = yx = 1$. Otherwise, x is called *singular*. If y exists, y is unique and it is usually denoted by x^{-1} and called the *inverse* of x.

If A has no unit element, then $x \in A$ is said to be *regular* if there exists $y \in A$ such that $x + y + xy = 0$. If y exists it is unique and is denoted by x^0, the *adverse* of x.

Remark. If a normed algebra A does not have a unit element, it is possible to embed A in a minimal superalgebra in which there exists a unit element. For instance, if A is a normed algebra without a unit, let $\tilde{A} = A \oplus K$, the direct sum of A and the scalar field K. Let (x, α) denote an element of $\tilde{A}$, where $x \in A$ and $\alpha \in K$. Clearly, $\tilde{A}$ is a linear space. With the norm: $\|(x, \alpha)\| = \|x\| + |\alpha|$, and the product: $(x, \alpha)(y, \beta) = (\alpha y + \beta x + xy, \alpha\beta)$, $\tilde{A}$ becomes a normed algebra with identity $(0, 1)$. The mapping $x \to (x, 0)$ embeds A into $\tilde{A}$ isometrically. Thus, the regularity of an element in a normed algebra without a unit can be defined in terms of the regularity of elements in a normed algebra with a unit element. We shall not deal with the technical details here. The interested reader can consult Rickart[50] for instance. Moreover, we shall assume that $\|1\| = 1$ whenever the algebra A has identity 1.

If A is not commutative, then one defines left and right regular elements. Some of the properties of regular elements are contained in the following:

Proposition 1. *Let A be a Banach algebra with a unit element. Then:*

(a) *For each $x \in A$ such that $\|x - 1\| < 1$, x is regular and its inverse x^{-1} is given by*

$$x^{-1} = 1 + \sum_{n=1}^{\infty} (1 - x)^n.$$

(*If A has no unit element and if $\|x\| < 1$, then x has an adverse x^0 given by $x^0 = -\sum_{n=1}^{\infty} x^n$ and satisfying $xx^0 - x - x^0 = x^0 x - x - x^0 = 0$.*)

(b) *The set G_r of all regular elements in A forms a multiplicative group.*

(c) *For all complex $\lambda \neq 0$, $\lambda G_r \subset G_r$.*

(d) *The group G_r is an open subset of A.*

(e) *The group G_r is a Hausdorff topological group.*

(f) *The mapping $x \to x^{-1}$ of G_r onto itself is a homeomorphism.*

PROOF. (a) By hypothesis,

$$\|(1 - x)^n\| \leq \|1 - x\|^n < 1.$$

By completeness of A, $y = 1 + \sum_{n=1}^{\infty} (1 - x)^n$ exists, since the series converges absolutely in A. By the joint continuity of the product in A, we have

$$yx = x + \sum_{n=1}^{\infty}(1-x)^n x = x - \sum_{n=1}^{\infty}(1-x)^n(1-x) + \sum_{n=1}^{\infty}(1-x)^n$$

$$= x - \sum_{n=1}^{\infty}(1-x)^{n+1} + \sum_{n=1}^{\infty}(1-x)^n = x + (1-x) = 1.$$

Similarly, $xy = 1$. Hence, $y = x^{-1} = 1 + \sum_{n=1}^{\infty}(1-x)^n$. (The other part follows similarly by showing that $\sum_{n=1}^{\infty} x^n$ is absolutely convergent.)

(b) Let x, $y \in G_r$. Then it is clear that $xy \in G_r$, since $(xy)(y^{-1}x^{-1}) = 1$ by associativity of the product. Clearly, $1 \in G_r$. Moreover, for each $x \in G_r$, $x^{-1} \in G_r$, because the inverse of x^{-1} is x.

(c) The proof is clear.

(d) Let $x_0 \in G_r$. Let $x \in A$ such that $\|x - x_0\| < \|x_0^{-1}\|^{-1}$. Since $\|x_0^{-1}x - 1\| = \|x_0^{-1}(x - x_0)\| \le \|x_0^{-1}\| \|x - x_0\| < 1$, by (a) $x_0^{-1}x \in G_r$ and hence by (b), $x = x_0(x_0^{-1}x) \in G_r$. This proves that G_r is an open subset of A.

(e) Since the product in A is jointly continuous, it is jointly continuous in G_r. Now for (e) it is sufficient to show that the mapping $x \to x^{-1}$ of G_r onto itself is continuous. Let $\varepsilon > 0$ be given. We wish to find $\delta > 0$ such that $\|x^{-1} - x_0^{-1}\| < \varepsilon$, whenever $\|x - x_0\| < \delta$, x, $x_0 \in G_r$. In view of (d), let $x \in G_r$ such that $\|x - x_0\| < 1/2 \|x_0^{-1}\|$. Since $x_0^{-1}x \in G$ (by (b)), by (a) we have

$$x^{-1}x_0 = (x_0^{-1}x)^{-1} = 1 + \sum_{n=1}^{\infty}(1-x_0^{-1}x)^n.$$

Also $\|x_0^{-1}x - 1\| \le \|x_0^{-1}\| \|x - x_0\| < 1/2$. But then

$$\|x^{-1} - x_0^{-1}\| = \|(x^{-1}x_0 - 1)x_0^{-1}\| \le \|x_0^{-1}\| \|x^{-1}x_0 - 1\|$$

$$\le \|x_0^{-1}\| \sum_{n=1}^{\infty}\|1 - x_0^{-1}x\|^n$$

$$\le \|x_0^{-1}\| \|1 - x_0^{-1}x\|/(1 - \|1 - x_0^{-1}x\|)$$

$$\le 2 \|x_0^{-1}\| \|1 - x_0^{-1}x\| \le 2 \|x_0^{-1}\|^2 \|x - x_0\|.$$

Now choose $\delta = \varepsilon/2 \|x_0^{-1}\|^2$. Then for $\|x - x_0\| < \delta$, we have

$$\|x^{-1} - x_0^{-1}\| < 2 \|x_0^{-1}\|^2 \delta = \varepsilon.$$

(f) This result follows from (e), in view of Theorem 2, §19, Chapter III.

Let A be a Banach algebra with an identity. Let $x \in A$. The set of all complex numbers λ such that $x - \lambda 1$ is not a regular element of A is called the *spectrum* of x, and is denoted by $Sp(x)$.

If A does not have an identity, we define the spectrum of an element $x \in A$ to be the set of complex numbers that becomes the spectrum of x when A is enlarged by adding an identity. In this case, $\lambda = 0$ is always in

$Sp(x)$ for any $x \in A$. Indeed, an element $x \in A$ cannot have an inverse in the extended algebra $A \oplus C$ because otherwise the identity of $A \oplus C$ would be in A. Furthermore, in this case $\lambda \neq 0$ is in $Sp(x)$ if, and only if, x/λ does not have an adverse.

Proposition 2. (a) *The spectrum $Sp(x)$ is a compact subset of the complex plane.*

(b) *The function $g : \lambda \to (x - \lambda 1)^{-1}$, defined on the complement $Sp'(x)$ of $Sp(x)$, is continuous.*

(c) *For $\lambda, \mu \in Sp'(x)$,*

$$g(\lambda) - g(\mu) = (\lambda - \mu)g(\lambda)g(\mu).$$

(d) *The set $Sp(x)$ is not empty.*

PROOF. (a) Since the mapping $\lambda \to x - \lambda 1$ of the complex plane into A is trivially continuous, and since the set of all singular elements is closed (because G_r is open), $Sp(x)$, being the inverse image of the closed set under the continuous mapping, is closed. Thus, to establish (a), it is sufficient to show that $Sp(x)$ is bounded in the complex plane. If not, there exists $\lambda \in Sp(x)$ such that $|\lambda| > \|x\|$, or $\|\lambda^{-1}x\| < 1$. But then $\|1 - (1 - \lambda^{-1}x)\| < 1$ implies that $1 - \lambda^{-1}x$ is regular by Proposition 1(a). Hence by Proposition 1(c), $(1 - \lambda^{-1}x)(-\lambda) = x - \lambda 1 \in G_r$, which is a contradiction.

(b) For $\lambda \in Sp'(x)$, $g(\lambda) = (x - \lambda 1)^{-1} \in G_r$.
Let $\lambda_0 \in Sp'(x)$. Then

$$\|g(\lambda) - g(\lambda_0)\| = \|(x - \lambda 1)^{-1} - (x - \lambda_0 1)^{-1}\| \to 0$$

as $\lambda \to \lambda_0$, owing to (f) in Proposition 1.

(c) For $\lambda, \mu \in Sp'(x)$, $g(\lambda) = (x - \lambda 1)^{-1}$ implies

$$\begin{aligned}
g(\lambda) &= g(\lambda)(x - \mu 1)g(\mu) \\
&= g(\lambda)[x - \lambda 1 + (\lambda - \mu)1]g(\mu) \\
&= [g(\lambda)(x - \lambda 1) + (\lambda - \mu)g(\lambda)1]g(\mu) \\
&= [1 + (\lambda - \mu)g(\lambda)]g(\mu).
\end{aligned}$$

Hence,

$$g(\lambda) - g(\mu) = (\lambda - \mu)g(\lambda)g(\mu).$$

(d) Let A' denote the dual of A, i.e., A' is the set of all linear continuous complex-valued mappings of A. The set A' is a topological linear space with the topology defined by the norm $\|f\|_\infty = \sup_{\|x\| \leq 1} |f(x)|$, $f \in A'$, $x \in A$. Let f be an arbitrary element of A'. Consider the composition mapping $f \circ g$ of g (defined in (b)) and f. Since f and g are continuous, $f \circ g$ is a continuous complex-valued function defined on $Sp'(x)$. First we show that it is analytic. Owing to linearity of f, by (c), we have

$$f(g(\lambda)) - f(g(\lambda_0)) = (\lambda - \lambda_0)f(g(\lambda)g(\lambda_0)), \quad \lambda, \lambda_0 \in Sp'(x).$$

This implies that

$$\lim_{\lambda \to \lambda_0} [\{f \circ g(\lambda) - f \circ g(\lambda_0)\}/(\lambda - \lambda_0)] = f([g(\lambda_0)]^2).$$

Hence, $f \circ g$ is analytic at each $\lambda_0 \in Sp'(x)$.

Clearly, $g(\lambda) = \lambda^{-1}(x\lambda^{-1} - 1)^{-1} \to 0$ as $\lambda \to \infty$, since $(x\lambda^{-1} - 1)^{-1} \to -1$ as $\lambda \to \infty$ by Proposition 1(a). But

$$|f \circ g(\lambda)| \le \|f\|_\infty \|g(\lambda)\|$$

implies that $f \circ g(\lambda) \to 0$ as $\lambda \to \infty$. This shows that $f \circ g$ is bounded outside a circle of sufficiently large radius with the origin of the complex plane as its center.

Now suppose $Sp(x)$ is empty. Then $Sp'(x)$ is the whole complex plane. Moreover, $f \circ g$ is analytic on the whole complex plane and, therefore, bounded on any closed bounded subset of the complex plane. Since $f \circ g$ has been shown to be bounded outside a bounded region, it is bounded everywhere. Hence, by Liouville's theorem (Titchmarsh,[44] p. 85), $f \circ g$ is a constant and thus $f \circ g(\lambda) = f(g(\lambda)) = 0$ for all complex λ. But f, being an arbitrary element in A', from the duality between A and A', it follows that $g(\lambda) = 0$ for all complex λ. But this is impossible; for the inverse of any element in A can never be zero. Hence, $Sp(x)$ cannot be empty. This completes the proof.

Part (d) of the previous proposition leads us to the famous theorem of Gelfand and Mazur. First we recall (Van der Waerden[45]) that a *division algebra* A is an algebra with an identity and such that for a, $b \in A$, $b \ne 0$, there exists $c \in A$ such that $a = bc$. The element c is usually written as a/b. This implies that each nonzero $x \in A$ has an inverse. Conversely, if each element in an algebra has an inverse then it is a division algebra.

Theorem 1 (*Gelfand-Mazur*). *Every Banach division algebra A is isomorphic and homeomorphic with the complex numbers.*

PROOF. We shall show that for each $x \in A$ there exists a complex number λ such that $x = \lambda 1$, in which 1 is the identity of A. This fact implies that the mapping $x \to \lambda 1$ is a norm-preserving isomorphism. Since $\lambda \to \lambda 1$ is an isomorphism of the complex numbers on the set $\{\lambda 1\}$, λ complex, it immediately follows, by combining the two isomorphisms, that A is isometrically isomorphic with the complex numbers.

Now to show that for each $x \in A$, there exists λ such that $x = \lambda 1$, assume that for some x no such λ exists. Then for each complex λ, $x - \lambda 1 \ne 0$ and, hence, $(x - \lambda 1)^{-1}$ exists for each λ. That means $x - \lambda 1$ is regular for each λ. That means that $Sp(x)$ is empty, which contradicts (d) of Proposition 2. This completes the proof.

The positive number defined by

$$r(x) = \sup \{|\lambda| : \lambda \in Sp(x)\}$$

is called the *spectral radius*. We have the following:

$$r(x) = \lim_{n \to \infty} \|x^n\|^{1/n}.$$

To establish this, we first observe that for any $x \in A$, when A has an identity, $x^n - \lambda 1 = \prod_{i=1}^{n} (x - \lambda_i 1)$, in which λ_i's are n roots of the polynomial $t^n - \lambda$. Hence, $x^n - \lambda 1$ is singular if, and only if, $x - \lambda_i 1$ is singular for at least one i. This shows that $Sp(x^n) = [Sp(x)]^n$ and, hence, $r(x^n) = [r(x)]^n$. Since $r(x^n) \le \|x^n\|$, $[r(x)]^n \le \|x^n\|$. Or $r(x) \le \|x^n\|^{n^{-1}} \le \|x\|$. To prove the reverse inequality, it is sufficient to show that for any real number a, $r(x) < a$ implies $\|x^n\|^{n^{-1}} \le a$ for all but a finite number of n's. Since for $\lambda \in Sp(x)$, $|\lambda| \le \|x\|$ (because $r(x) \le \|x\|$), for $|\lambda| > \|x\|$ we have

$$g(\lambda) = (x - \lambda 1)^{-1} = -\lambda^{-1}\left(1 + \sum_{n=1}^{\infty} x^n \lambda^{-n}\right)$$

by Proposition 1(a). Now if f is any linear functional in A' (the dual of A) then

(*)
$$f \circ g(\lambda) = f(g(\lambda)) = -\lambda^{-1}\left(f(1) + \sum_{n=1}^{\infty} f(x^n)\lambda^{-n}\right)$$

for $|\lambda| > \|x\|$. Since $f \circ g$ is analytic (cf., proof of (d) in Proposition 2) in $Sp'(x)$, the Laurent expansion of $f \circ g$ given in (*) is valid for $|\lambda| > r(x)$, as is known from complex analysis. Hence, for any real b, $r(x) < b < a$, the series $\sum_{n=1}^{\infty} f(x^n)b^{-n}$ is convergent and, therefore, the sequence $\{f(x^n)b^{-n}\}$ is bounded for each $f \in A'$. Since in a Banach space weakly bounded and norm bounded sets are the same (cf., Husain,[22] p. 30, Theorem 8), $\|x^n b^{-n}\| \le M$, for some $M > 0$, and for all $n \ge 1$. Thus, $\|x^n\|^{n^{-1}} \le M^{n^{-1}}b$ for all $n \ge 1$ and for sufficiently large n, $M^{n^{-1}}b \le a$. This proves that $\|x^n\|^{n^{-1}} \le a$ for sufficiently large n and the formula is established. If A does not have an identity, one enlarges A to an algebra with an identity, and the same proof works.

Since $\|x^n\| \le \|x\|^n$ by the definition of a Banach algebra, it follows that $r(x) \le \|x\|$. The equality $r(x) = \|x\|$ holds if, and only if, $\|x^2\| = \|x\|^2$. For, if $\|x^2\| = \|x\|^2$ then $\|x^{2^n}\| = \|x\|^{2^n}$ and, hence, $r(x) = \lim_{n \to \infty} \|x^n\|^{n^{-1}} = \lim_{n \to \infty} \|x^{2^n}\|^{2^{-n}} = \lim_{n \to \infty} \|x\| = \|x\|$. Conversely, if $r(x) = \|x\|$, then $\|x^2\| = r(x^2) = [r(x)]^2 = \|x\|^2$.

Although in general Banach algebras it is not true that $\|x^2\| = \|x\|^2$, in some special Banach algebras for some special elements x this is always true. Hence, we consider the following:

An operation "*" in an algebra A is called an *involution* if for each $x \in A$ there exists an element $x^* \in A$ and where:

(a) $(\alpha x + \beta y)^* = \bar{\alpha}x^* + \bar{\beta}y^*$, α, β complex and $\bar{\alpha}$, $\bar{\beta}$ their complex conjugates;

(b) $(xy)^* = y^* x^*$;

(c) $x^{**} = x$.

An algebra with an involution is called a *-*algebra*. An isomorphism f between two *-algebras is said to be a *-*isomorphism* if $f(x^*) = [f(x)]^*$. The element x^* is called the *adjoint* of x. If $x = x^*$, x is called *self-adjoint*. If $xx^* = x^*x$, x is called *normal*. A Banach *-algebra A is called a *B*-algebra* if

$$\|xx^*\| = \|x\|^2$$

for each $x \in A$. It is easy to check that in a B^*-algebra, $\|x^*\| = \|x\|$ and $\|x^*x\| = \|x^*\| \, \|x\|$.

Let A be a B^*-algebra and $x \in A$. Suppose x is normal. Then

$$
\begin{aligned}
\|x^*\|^2 \|x\|^2 &= (\|x^*\| \, \|x\|)^2 = (\|x^*x\|)^2 \\
&= \|(x^*x)^*(x^*x)\| = \|(x^*x^{**})(x^*x)\| \\
&= \|(x^*)^2 x^2\| = \|(x^2)^*(x^*x)\| \\
&= \|(x^*)^2 x^2\| = \|(x^2)^* x^2\| = \|(x^2)^*\| \, \|x^2\| \\
&\le \|x^*\|^2 \, \|x^2\|.
\end{aligned}
$$

Hence, $\|x\|^2 \le \|x^2\|$. But $\|x^2\| \le \|x\|^2$ in any normed algebra. Hence, for each normal $x \in A$, $\|x^2\| = \|x\|^2$. Therefore, by the above formula for $r(x)$, for each normal x in a B^*-algebra,

$$r(x) = \|x\|.$$

The notion of involution is available in the Banach algebras with which we are concerned in this book. For instance, if $f^*(x) = \overline{f(-x)}$, for each $f \in L_1(G)$, where G is a Hausdorff locally compact abelian topological group and $x \in G$, then $f \to f^*$ is an involution in $L_1(G)$. Also, we have seen that $\|f^*\|_1 = \|f\|_1$. (Observe $f^* = \tilde{f}$ in our earlier notation adopted in §47, Chapter VIII.) Since $L_1(G)$ is commutative if G is so, each $f \in L_1(G)$ is normal.

50. MAXIMAL IDEAL SPACE AND GELFAND-NAIMARK THEOREM

Since in algebras there is ring operation (multiplication), in addition to group operation (addition), it is natural that ideals play a significant role.

Let A be an algebra with a unit element. A subalgebra I of A is said to be a *left, right* or *two-sided* ideal of A if $AI \subset I$, $IA \subset I$ or both, respectively. Clearly, A and $\{0\}$ are two-sided ideals. They are called *trivial* ideals. In a commutative algebra each ideal is two-sided. A nontrivial ideal I is said to be *maximal* if there is no ideal (different from A) that properly contains I. By Zorn's lemma it can be shown that in an algebra with an identity each left or right ideal is contained in a left or right maximal ideal. If an algebra does not have an identity, the preceding statement is not true. However, one can rescue the situation by introducing the following notions:

Let A be an algebra. An ideal I of A is said to be a *left* (or *right*) *regular* ideal if I is a left (or right) ideal and if there is a right (or left) identity modulo

I. Now again by Zorn's lemma it can be shown that each left (or right) proper regular ideal is contained in a left (or right) regular maximal ideal.

The intersection of all left (or right) maximal ideals in the case when the algebra has a unit element, or the intersection of all left (or right) regular maximal ideals in the case when the algebra does not have a unit element, is said to be its *radical*. It can be shown that the radical of an algebra is always a two-sided ideal. If the radical of A is $\{0\}$, the algebra is said to be *semisimple*.

Let A be a Banach algebra. Then every regular maximal ideal M is a closed ideal: if u is an identity modulo M then for all x in M $\|u - x\| \geq 1$ since otherwise $(u - x)^0$ exists and then $u \in M$, a contradiction. If A has an identity then every maximal ideal is closed. Thus, the radical of a Banach algebra is a two-sided closed ideal.

As for linear spaces, the quotient algebra of a normed (or Banach) algebra A by a closed two-sided ideal I is a normed (or Banach) algebra. The quotient algebra of a Banach algebra by its radical is semisimple.

Recall that the norm of the quotient algebra A/I is defined by

$$\|\dot{x}\| = \|x + I\| = \inf_{i \in I} \|x + i\|,$$

where $\dot{x} = x + I \in A/I$. As usual, the mapping $x \to \dot{x}$ of A onto A/I is called the natural or canonical mapping. The canoncial mapping is an algebra-homomorphism of A onto A/I.

Let C be a commutative ring (or a commutative algebra) with an identity. Then C is a field if, and only if, C has no nontrivial ideals. Indeed, it is well known that a field cannot have any nontrivial ideals (Van der Waerden,[45] p. 53, Ex. 9). To show the converse, we must show that each nonzero x has an inverse. The set $I = \{xy : y \in C\}$ is clearly an ideal. Since $x \neq 0$ and $x \in C$ (because C has an identity), I is not the zero ideal. Therefore, $I = C$. This shows that $xy = e$ for some $y \in C$, where e is the identity of C. This shows that x has an inverse and so C is a field. We use this result below.

Proposition 3. *If A is a commutative Banach algebra and M a regular maximal ideal in A, then A/M is homeomorphic with the algebra of complex numbers.*

PROOF. Since M is a closed regular ideal, A/M is a Banach algebra with an identity. Also, A/M is commutative, since A is. Since M is a regular maximal ideal, A/M does not have any nontrivial ideals. Hence, A/M is a field by the above discussion. (Observe that each algebra is a ring.) Therefore, A/M being a field and thus a division algebra, by Theorem 1, §49, it is homeomorphically isomorphic to the algebra of complex numbers.

Remark. If the algebra A in Proposition 3 has, in addition, an identity, then the proposition remains true for any maximal ideal M instead of a regular maximal ideal M.

Proposition 3 is a very useful result in Banach algebra theory as we now show.

Let A be a commutative Banach algebra. For each regular maximal ideal M in A, let

$$\hat{x}(x, M) = x + M \in A/M.$$

Since, by Proposition 3, A/M is the algebra of the complex numbers, $\hat{x}$ is a complex-valued function of two variables defined on the product of A and $\mathfrak{M}_A$, the set of all regular maximal ideals in A.

For a fixed $x \in A$, $\hat{x}(x,.)$ is a complex-valued function on $\mathfrak{M}_A$. The mapping $x \to \hat{x}(x,.)$ of A into the set of all complex-valued functions on the set $\mathfrak{M}_A$ of all regular maximal ideals is better known as the *Gelfand mapping*.

We can endow $\mathfrak{M}_A$ with the coarsest topology with respect to which each $\hat{x}$ is continuous in $M \in \mathfrak{M}_A$ for fixed x. The set $\mathfrak{M}_A$, endowed with this topology, is called the *maximal ideal space*. If A has an identity, $\mathfrak{M}_A$ is the set of all maximal ideals.

To bring out part of the significance of the maximal ideal space, we first of all observe that for each $x \in A$, $\hat{x}$ is bounded on $\mathfrak{M}_A$. For,

$$|\hat{x}(x, M)| = |x + M| = \|x + M\| = \inf_{m \in M} \|x + m\| \le \|x\|,$$

since $0 \in M$. Hence,

$$\|\hat{x}(x, .)\|_\infty = \sup_{M \in \mathfrak{M}_A} |\hat{x}(x, M)| \le \|x\|.$$

Since $\hat{x}(x,.)$ is continuous by definition, it follows that $\hat{x}(x,.) \in C(\mathfrak{M}_A)$, the space of all continuous bounded complex-valued functions on $\mathfrak{M}_A$.

Furthermore, in view of the canonical homomorphism of A onto A/M it is clear that the Gelfand mapping $x \to \hat{x}(x,.)$ is a homomorphism of A into $C(\mathfrak{M}_A)$.

Thus we have shown the following:

Proposition 4. *Let A be a commutative Banach algebra. Then the Gelfand mapping $x \to \hat{x}(x,.)$ of A into $C(\mathfrak{M}_A)$ is a norm-decreasing (therefore continuous) homomorphism of A into $C(\mathfrak{M}_A)$.*

Proposition 4 established a representation for any commutative Banach algebra into a function algebra, i.e., the algebra of continuous bounded complex-valued functions on a topological space $\mathfrak{M}_A$ obtained from the given Banach algebra. The following proposition gives a criterion under which this representation is faithful (i.e., one-to-one).

Proposition 5. *The mapping $x \to \hat{x}(x,.)$ of a commutative Banach algebra A into $C(\mathfrak{M}_A)$ is one-to-one if, and only if, A is semisimple.*

PROOF. Observe that the radical of A is the set of all x for which $\hat{x}(x, M) = 0$ for all $M \in \mathfrak{M}_A$. Now, x is in the kernel of $x \to \hat{x}(x,.)$ if, and only if, $\hat{x}(x, M) = 0$ for all $M \in \mathfrak{M}_A$. Hence, the kernel of $x \to \hat{x}(x, .)$ is zero if, and only if, the radical is $\{0\}$, i.e., if, and only if, A is semisimple.

Proposition 5 states a criterion under which the Gelfand representation is faithful. The representation is not always onto $C(\mathfrak{M}_A)$.

In Propositions 4 and 5, we emphasized the continuity and boundedness of $\hat{x}(x, .)$ as a function in $C(\mathfrak{M}_A)$. Now we wish to consider the role of $\hat{x}$ when it is regarded as a function of x while M is kept fixed.

For each $M \in \mathfrak{M}_A$, $\hat{x}(x, M)$ is a multiplicative linear functional on A, since the canonical mapping of A into A/M is an algebra-homomorphism. Although $\hat{x}$, as a function of $M \in \mathfrak{M}_A$ for each fixed $x \in A$, has a unique preimage under restricted conditions, i.e., if A is semisimple, the fact that $\hat{x}$ is the unique image under the mapping $M \to \hat{x}(x, M)$ is generally true. More precisely we have the following:

Proposition 6. *For a fixed $x \in A$, the mapping $M \to \hat{x}(x, M)$ of $\mathfrak{M}_A$ onto the set of all nonzero linear continuous multiplicative functionals on a commutative Banach algebra A is one-to-one.*

PROOF.   If $M_1 \neq M_2$, then there exist $x \in M_1$ and $x \notin M_2$. But this implies that $\hat{x}(x, M_1) = 0$ and $\hat{x}(x, M_1) \neq 0$. Hence, the mapping $M \to \hat{x}(x, M)$ is one-to-one. To show that it is onto, let f be a nonzero continuous multiplicative linear functional on A. Then the kernel $f^{-1}(0) = M_0$ is a maximal ideal in A because the image of f is a field. Since the set of complex numbers admits no complex automorphism except the identity, two nonzero continuous multiplicative linear functionals coincide on A if they have the same kernel. Using this fact, we see that the mapping $M \to \hat{x}(x, M)$ is onto.

In view of the above proposition, we see that the maximal ideal space can be identified with the set of all nonzero continuous multiplicative linear functionals on A or, in other words, with a subset of the dual A' (the set of all continuous linear functionals on A) of A.

Actually, the maximal ideal space can be identified with the subset of the unit ball in A'. For, since

$$|\hat{x}(x, M)| = |x + M| \leq \|x\|,$$

we have

$$\|\hat{x}(., M)\|_\infty = \sup_{\|x\| \leq 1} |\hat{x}(x, M)| \leq 1.$$

Furthermore, since the weak* topology $\sigma(A', A)$ on A' (cf., §11, Chapter I) is the coarsest topology for which each linear functional is continuous, it is easy to see that the induced weak* topology on $\mathfrak{M}_A$, when $\mathfrak{M}_A$ is identified with the subset of the unit ball in A', coincides with the topology on $\mathfrak{M}_A$. Let Δ denote the image of $\mathfrak{M}_A$ in A' under the mapping $M \to \hat{x}(x, M)$.

Proposition 7.   *Let A be a commutative Banach algebra. Then:*
(a) *Δ is a relatively compact subset of A'.*
(b) *The maximal ideal space $\mathfrak{M}_A$ is a locally compact Hausdorff space.*
(c) *If A has an identity, $\mathfrak{M}_A$ is a compact Hausdorff space.*

PROOF. (a) Since the unit ball $S = \{x \in A : \|x\| \leq 1\}$ is a closed neighborhood of 0 in A, the polar or the unit ball $S' = \{x' \in A' : |\langle x, x' \rangle| \leq 1$ for all $x \in S\}$ in A' is an equicontinuous subset of A' (Husain,[22] Chapter II, §9, Proposition 14). But each equicontinuous subset is relatively weakly compact. Since S' is $\sigma(A', A)$-closed, it follows that S' is $\sigma(A', A)$-compact. But then Δ, being a subset of S', is relatively $\sigma(A', A)$-compact.

(b) Since $\sigma(A', A)$ on A' is a Hausdorff topology (Husain,[22] Chapter II) and since the topology of $\mathfrak{M}_A$ coincides with $\sigma(A', A)$ when $\mathfrak{M}_A$ is identified with Δ, it follows that $\mathfrak{M}_A$ is a Hausdorff space. Let H denote the set of all multiplicative linear functionals in S'. Clearly, if C is the set of complex numbers then

$$H = \bigcap_{x, y \in A} \{f \in S' : f(xy) - f(x)f(y) = 0\} \subset C^A.$$

Since the projection mapping $C^A \to C$ is continuous and $\{0\}$ is a closed subset of C, it follows that H is a closed subset of S'. (Observe that the product topology on C^A induces $\sigma(A', A)$ on A'.) But by Proposition 6 and the remarks following it, $H = \Delta \cup \{0\}$. Since S' is $\sigma(A', A)$-compact (see (a)) and H is a closed subset of S', H is also $\sigma(A', A)$-compact. This shows that H is the one-point compactification of Δ. Hence, Δ is a locally compact Hausdorff space.

(c) Assume A has an identity. In view of a well-known fact about one-point compactification (cf., Kelley,[27] p. 150), to show that Δ or $\mathfrak{M}_A$ is compact it is sufficient to show that the zero homomorphism is an isolated point (i.e., $\{0\}$ is both closed and open). Since $\{0\}$ is a singleton in a Hausdorff space, $\{0\}$ is closed. To show that $\{0\}$ is open, we show that 0 cannot be a limit point of Δ. Let e denote the identity of A. Then for each $h \in \Delta$, $h(e) = 1$. But the zero homomorphism maps A onto the zero of the complex number. If 0 were a limit point of Δ, we would have $0 = 1$, which is impossible. Hence, 0 is an isolated point. Therefore, Δ is a compact Hausdorff space.

Remark. $\{0\}$ is the "point at infinity" in the one-point compactification of $\mathfrak{M}_A$ or Δ, and we may denote $\{0\}$ by "∞" in order to conform to the general practice.

Once we have identified $\mathfrak{M}_A$ with $\Delta \subset A'$, it is of interest to locate $\hat{x}$. It is quite clear that $\hat{x}$ is an element of $C^{A'}$, the set of all complex-valued mappings of A'. As we saw before, $\hat{x}$ is actually a bounded continuous function on $\mathfrak{M}_A$ or Δ. Now take an element x'' of the bidual (§11, Chapter I) A'' of A, or equivalently the dual of A', and restrict its domain to Δ. (Observe that the domain of x'' is the whole space A'.) Then the restriction $x'' | \Delta$ is precisely $\hat{x} = \hat{x}(x, M)$. For, if $x'_M \in \mathfrak{M}_A$, then $x''(x'_M) = x'_M(x) = x + M = \hat{x}(x, M)$. Let $\hat{A} = \{\hat{x} = \hat{x}(x, .) : x \in A\}$.

Corollary 1. *Let A be a commutative Banach algebra without an identity. Then for a fixed $x \in A$, each $\hat{x} = \hat{x}(x, M) \in C(\mathfrak{M}_A)$ vanishes at ∞.*

PROOF. Since each $\hat{x} = \hat{x}(x, M)$ is continuous on $\mathfrak{M}_A$ or Δ, $\hat{x}$ can be extended continuously on the one-point compactification, as is well-known (cf., §11, Chapter I). But since the zero homomorphism or ∞ is a limit point of Δ or $\mathfrak{M}_A$ and since a zero homomorphism maps each element of A onto the zero of the complex number, the corollary follows.

In view of Corollary 1 and the remarks preceding it, we see that $\hat{A} \subset C_\infty(\mathfrak{M}_A) \subset C(\mathfrak{M}_A)$. As remarked after Proposition 5, the mapping $x \to \hat{x}(x,.)$ is not always onto $C_\infty(\mathfrak{M}_A)$. We have the following:

Proposition 8. *For each $x \in A$, put $\hat{x} = \hat{x}(x,.)$. Let $\hat{A}$ denote the set of all functions $\hat{x}$ as x runs over a commutative Banach algebra A. Then:*

(a) *$\hat{A}$ is a normed algebra with the sup norm $\|\hat{x}\|_\infty = \sup\limits_{M \in \mathfrak{M}_A} |\hat{x}(x, M)|$, and hence a subalgebra of $C_\infty(\mathfrak{M}_A)$.*

(b) *$r(x) = \|\hat{x}\|_\infty$.*

(c) *$\hat{A}$ is dense in $C_\infty(\mathfrak{M}_A)$ if for each $x \in A$ there exists a $y \in A$ such that $\hat{y} = \bar{\hat{x}}$.*

(d) *$\hat{A} = C_\infty(\mathfrak{M}_A)$ if for each $x \in A$ there exists a $y \in A$ such that $\hat{y} = \bar{\hat{x}}$ and for each $x \in A$, $\|x^2\| = \|x\|^2$.*

PROOF.   The proof of (a) is simple.

(b) Recall that $r(x) = \sup\{|\lambda| : \lambda \in Sp(x)\}$. First of all, we show that the range of $\hat{x}$ is identical with either the $Sp(x)$ or with $Sp(x) \sim \{0\}$. If $\lambda \neq 0$, then $\lambda \in Sp(x)$ if, and only if, x/λ does not have an adverse or, in other words, x/λ is the relative identity for some regular maximal ideal M. Let f_M be the homomorphism of A determined as follows: Since A/M is the field, let $\dot{e}$ denote its identity. For each coset $\dot{x} = x + M \in A/M$ containing x, there is a complex number λ such that $\dot{x} = \lambda\dot{e}$. Define $f_M(x) = \lambda$. Then it is easy to check that f_M is a complex homomorphism of A whose kernel is M. It is clear that x/λ is a relative identity for M if, and only if, $f_M(x/\lambda) = 1$ or $f_M(x) = \lambda$. But $f_M(x) = \hat{x}(x, M)$. Hence, $\hat{x}(x, M) = \lambda$. If $f_M(x) = 0$, then $x \in M$. Hence $0 \in Sp(x)$. This proves that the range of $\hat{x}$, for a fixed x, coincides with the $Sp(x)$ or $Sp(x) \sim \{0\}$. Hence, $\|\hat{x}\|_\infty = r(x)$.

(c) The proof of (c) depends upon the Stone-Weierstrass theorem (Theorem 7, §11, Chapter I). By (a), $\hat{A}$ is a subalgebra of $C_\infty(\mathfrak{M}_A)$. Moreover, if $M_1 \neq M_2$, $M_1, M_2 \in \mathfrak{M}_A$, then there is $x \in M_1$ such that $x \notin M_2$. This shows that $\hat{x}(x, M_1) = 0$ and $\hat{x}(x, M_2) \neq 0$. This means that the sub-algebra $\hat{A}$ separates points. Now from the hypothesis combined with this property, it follows that $\hat{A}$ is dense in $C_\infty(\mathfrak{M}_A)$, by the Stone-Weierstrass theorem (§11, Chapter I).

(d) First we show that $\|\hat{x}\|_\infty = \|x\|$ for each $x \in A$. By (b), $r(x) = \|\hat{x}\|_\infty$. As pointed out after Theorem 1, §49, $r(x) = \|x\|$ if, and only if, $\|x^2\| = \|x\|^2$. Hence, $\|\hat{x}\|_\infty = \|x\|$. This shows that the mapping $x \to \hat{x}$ is isometric. Since A is a Banach algebra, so is $\hat{A}$. Hence $\hat{A}$ is a closed subalgebra of $C_\infty(\mathfrak{M}_A)$. But in view of (c), it follows that $\hat{A} = C_\infty(\mathfrak{M}_A)$.

Remark.   (a) If the commutative Banach algebra A has an identity, then $\mathfrak{M}_A$ is a compact Hausdorff space (Proposition 7(c)). Hence, it is clear that $C_\infty(\mathfrak{M}_A) = C(\mathfrak{M}_A)$. Thus, in Proposition 8(b) and (c), we obtain a stronger conclusion if we replace $C_\infty(\mathfrak{M}_A)$ by $C(\mathfrak{M}_A)$.

(b) If the conditions of part (d) of Proposition 8 are satisfied, we can identify A with $\hat{A}$.

Now, returning to $A = L_1(G)$, where G is a locally compact Hausdorff abelian topological group, we note that A need not have an identity. Let $\mathfrak{M}_A$ denote the set of all regular maximal ideals of $L_1(G)$, or the maximal ideal space.

For each $M \in \mathfrak{M}_A$, A/M is a Banach division algebra and, hence homeomorphic with the algebra of complex numbers (Theorem 1, §49). For each $f \in L_1(G)$, $\hat{f} = \hat{f}(f,.)$ is a complex-valued function defined on $\mathfrak{M}_A$, which is a locally compact Hausdorff space.

For each $x \in G$, $M \in \mathfrak{M}_A$, and $f \in L_1(G) \sim M$, define $\hat{x}(x, M) = \hat{f}_x(f, M)/\hat{f}(f, M)$, where $f_x(y) = f(xy)$ for all $y \in G$ (i.e., f_x is the *left translate* of f). Clearly, $\hat{x}$ is a complex-valued function on the product $G \times \mathfrak{M}_A$. It can be shown that $\hat{x}$ is independent of the choice of $f \notin M$. Further,

(i) $\hat{x}(xy, M) = \hat{x}(x, M)\hat{x}(y, M)$

(ii) $|\hat{x}(x, M)| = 1$

(iii) $\hat{x}(x, M)$ is continuous in both variables together.

All these properties follow from the discussion before Proposition 4; (i) to (iii) state that for each fixed $M \in \mathfrak{M}_A$, $\hat{x}(., M)$ is a continuous character of G. Hence, $\hat{x}(., M) \in G'$ for each fixed $M \in \mathfrak{M}_A$. Put $x'_M = \hat{x}(., M)$. As in Proposition 6, the mapping $M \to x'_M$ of $\mathfrak{M}_A$ into G' is one-to-one, and onto. Hence, the maximal ideal space can be identified with the dual G' of G. Since $\mathfrak{M}_A$ is locally compact, G' can be identified with the locally compact topology induced from $\mathfrak{M}_A$. If we write $x'_M(x) = \langle x, x' \rangle$, we can show that G', endowed with this locally compact topology, is a topological group. Thus, one can deduce that the dual group G' of a locally compact Hausdorff abelian topological group is also locally compact under the topology described above, and G' is the maximal ideal space of $L_1(G)$.

So far we have seen how the basic notions of Fourier transform, involution, etc., in the study of $L_1(G)$ can be described abstractly in Banach algebras. In the remainder of this section we shall see how the spectral theorem, which played an important role in the representation theory of compact groups (Chapter VII), can be derived from a theorem in Banach algebras.

First we prove the following:

Theorem 2 (*Gelfand-Naimark*). *Let A be a commutative B*-algebra (§49, following Theorem 1) with an identity. Then the Gelfand mapping $x \to \hat{x}(x,.)$ is an isometric *-isomorphism of A onto the commutative B*-algebra $C(\mathfrak{M}_A)$, where $\mathfrak{M}_A$ is the maximal ideal space of A.*

PROOF. Observe that $\mathfrak{M}_A$ is a compact Hausdorff space, since A has an identity (Proposition 7). Furthermore, it is easy to verify that the function algebra $C(\mathfrak{M}_A)$, the space of all continuous complex functions on $\mathfrak{M}_A$, is a B^*-algebra with involution defined by $f^*(x) = \overline{f(x)}$ (complex conjugate of $f(x)$).

Since A is commutative, each element x of A is normal and, hence, $\|x^2\| = \|x\|^2$ (§49, following Theorem 1). To prove the theorem it is sufficient to show that $\hat{A} = C(\mathfrak{M}_A)$. But in view of Proposition 8(d), it is sufficient to show that $\hat{x}^* = \bar{\hat{x}}$, since $x^* \in A$ for each $x \in A$.

First we prove that if $x \in A$ is self-adjoint (i.e., $x^* = x$) then $\bar{\hat{x}} = \hat{x}$ (i.e., $\hat{x}$ is real). If not, then there exists an $M \in \mathfrak{M}_A$ such that $\hat{x}(x, M) = a + ib$, $b \neq 0$, a, b real. Putting $y = (x - a1)b^{-1}$, we have $y^* = y$, because $x^* = x$. Furthermore, $\hat{y}(y, M) = i$ and so $y - i1 \in M$. The image M^* of M under the involution in A is a maximal ideal, as is easy to check. Since $(y - i1)^* = y + i1 \in M^*$, $\hat{y}(y, M^*) = -i$. Moreover, for any $\lambda > 0$, if $z_1 = y - i\lambda 1$ and $z_2 = y + i\lambda 1$, then $\hat{z}_1(z_1, M^*) = -i(1 + \lambda)$ and $\hat{z}_2(z_2, M) = i(1 + \lambda)$. Hence $1 + \lambda \leq \|\hat{z}_1(z_1, M^*)\|_\infty \leq \|z_1\|$. Similarly $1 + \lambda \leq \|z_2\|$. But then

$$(1 + \lambda)^2 \leq \|z_1\| \, \|z_2\| = \|z_2^*\| \, \|z_2\| = \|z_2^* z_2\| = \|z_1 z_2\| = \|y^2 + \lambda^2 1\|$$
$$\leq \|y^2\| + \lambda^2,$$

or $1 + 2\lambda \leq \|y^2\|$. Since λ is arbitrary, this is impossible. Hence, $\hat{x}$ must be real.

Now let x be any element in A. Putting $y = 2^{-1}(x + x^*)$ and $z = (2i)^{-1}(x - x^*)$, we have $x = y + iz$. Clearly, y and z are self-adjoint. Hence, by the result of the previous paragraph, $\hat{y}$ and $\hat{z}$ are real, and so $\hat{x}^* = \widehat{(y - iz)} = \hat{y} - i\hat{z} = \bar{\hat{y}} - i\bar{\hat{z}} = \overline{\hat{y} + i\hat{z}} = \bar{\hat{x}}$. This proves the theorem.

A particular case of the above theorem is:

Corollary 2. (a) *Let A be a commutative C^*-algebra (§48, II) with an identity. Then $\hat{A} = C(\mathfrak{M}_A)$.*

(b) *Let T be a normal (i.e., $TT^* = T^*T$) operator on a Hilbert space H. Let A_T denote the closed subalgebra with an identity generated by T (i.e., A_T is the intersection of all closed subalgebras (of $B(H)$) with an identity and containing T and its complex conjugate operator T^*) (cf., Exercise 5, Chapter VII). Then $\hat{A} = C(\mathfrak{M}_{A_T})$.*

PROOF. (a) It is a particular case of Theorem 2, since each C^*-algebra is a B^*-algebra.

(b) The algebra A_T is easily seen to be the norm closure in $B(H)$ of all polynomials in T and T^*. Since T commutes with T^*, the algebra of all polynomials in T and T^* is commutative. Hence, its closure A_T is commutative. Thus, A_T is a commutative C^*-algebra with an identity, and (a) applies.

Before we prove the result that yields the spectral theorem, we need the following:

Proposition 9. *Let A be a commutative C^*-algebra of operators on a Hilbert space H. If an operator $T \in A$ is regular in $B(H)$, then $T^{-1} \in A$.*

PROOF. First assume that T is self-adjoint (i.e., $T = T^*$). Since T is regular, so is T^{-1}. Let $A_{T,T^{-1}}$ denote the closed subalgebra of $B(H)$ generated by T and T^{-1}. Clearly, $A_{T,T^{-1}}$ is a C^*-algebra. Since T commutes with T^{-1}, $A_{T,T^{-1}}$ is commutative. By Corollary 2(b), $A_{T,T^{-1}} = C(\mathfrak{M}_{A_{T,T^{-1}}})$, where $\mathfrak{M}_{A_{T,T^{-1}}}$ is the maximal ideal space of $A_{T,T^{-1}}$, a compact Hausdorff space. Thus, T can be identified with a real function in $C(\mathfrak{M}_{A_{T,T^{-1}}})$. Since $A \cap A_{T,T^{-1}} = B$ is a Banach subalgebra of $A_{T,T^{-1}}$, it follows that B is isomorphic with a subalgebra of $C(\mathfrak{M}_{A_{T,T^{-1}}})$. Since $T \in B$ is regular in $A_{T,T^{-1}}$, the proposition with the additional hypothesis that T is self-adjoint follows if we show that for each regular element f of a closed subalgebra A of $C(X)$, where X is a compact Hausdorff space, $f^{-1} \in A$. Since f is a continuous real function, $f(X)$ is a compact subset of the real numbers, and $f(x) \neq 0$ for all $x \in X$ because f is regular. Since the real function t^{-1} is continuous for all $t \neq 0$, in particular on $f(X)$, by the classical Weierstrass approximation theorem, t^{-1} can be approximated uniformly by real polynomials on $f(X)$. This shows that f^{-1} can be approximated by polynomials in f. Since A is a closed algebra, $f^{-1} \in A$.

To complete the proof, let T be any operator in A and regular in $B(H)$. Then $S = TT^*$ is self-adjoint and $S^{-1} = T^{*-1}T^{-1} \in A$ by the above particular case. Since A is commutative, $T(T^*S^{-1}) = (T^*S^{-1})T$ shows that $T * S^{-1} = T^{-1} \in A$.

Remark. It is quite clear from the above proposition that the spectrum of an operator $T \in A \subset B(H)$ coincides with the spectrum of T when T is regarded as an element of a commutative C^*-algebra A with an identity. In particular, if T is a normal operator in $B(H)$, the spectrum of T coincides with its spectrum as an element of the C^*-algebra generated by T and T^*.

Theorem 3. *Let T be a normal operator on a Hilbert space H. Let A_T be the commutative C^*-algebra with an identity generated by T (cf., Corollary 2(b)) and let $\mathfrak{M}_{A_T}$ be the maximal ideal space of A_T. Then the function $\hat{T} \in C(\mathfrak{M}_{A_T})$ maps $\mathfrak{M}_{A_T}$ homeomorphically onto $Sp(T)$, the spectrum of T.*

PROOF. As remarked above, $Sp(T)$ coincides with the spectrum of T as an operator on H. Furthermore, $\hat{T}$ is a continuous complex function on $\mathfrak{M}_{A_T}$ which is a compact Hausdorff space because A_T has an identity. By Proposition 2(a), §49, $Sp(T)$ is compact and, in view of the proof of Proposition 8(b), the range of $\hat{T}$ coincides with $Sp(T)$. Thus, $\hat{T}$ is a complex-valued continuous function with the compact domain $\mathfrak{M}_{A_T}$ and the compact range $Sp(T)$. To complete the proof, it is sufficient to show that $\hat{T}$ is one-to-one. If $M_1, M_2 \in \mathfrak{M}_{A_T}$ are such that $\hat{T}(T, M_1) = \hat{T}(T, M_2)$, $T \in A_T$, then

$$\hat{T}^*(T^*, M_1) = \overline{\hat{T}(T, M_1)} = \overline{\hat{T}(T, M_2)} = \hat{T}^*(T^*, M_2).$$

In other words, each of the functions $\hat{T}$ and $\hat{T}^*$ take equal values at M_1 and M_2. Since A_T is the closure of the set of all polynomials in T and T^* and since $A_T = C(\mathfrak{M}_{A_T})$, each function in $C(\mathfrak{M}_{A_U})$ is the uniform limit of polynomials in $\hat{T}$ and $\hat{T}^*$. Therefore, each polynomial, and hence, each function in $C(\mathfrak{M}_{A_T})$ takes equal values at M_1 and M_2. Now if $M_1 \neq M_2$, there exists a function $f \in C(\mathfrak{M}_{A_T})$ such that $f(M_1) \neq f(M_2)$, since $\mathfrak{M}_{A_T}$ being a compact Hausdorff space is completely regular. This is a contradiction and therefore $M_1 = M_2$. This completes the proof.

Observe that, in view of the above theorem, the maximal ideal space $\mathfrak{M}_{A_T}$ of the commutative C^*-algebra A_T generated by a normal operator T can be identified with the spectrum $Sp(T)$ of T. Thus, it is clear that $\hat{T}(T, M) = M$ for each $M \in \mathfrak{M}_{A_T}$.

A number of results can be derived from the above theorem. An important one is the spectral theorem.

Let T be a Hermitian or self-adjoint (hence, normal) operator on a Hilbert space H and $\hat{T}$ the corresponding continuous real function on the maximal ideal space $\mathfrak{M}_{A_T}$ of the closed C^*-algebra A_T with an identity generated by T in $B(H)$. Since $\mathfrak{M}_{A_T}$ is compact, $\hat{T}$ assumes its maximum and minimum values b and a, respectively. Let $a = \lambda_0 < \lambda_1 < \lambda_2 < \cdots < \lambda_n = b$ be a partition of the closed finite interval $[a, b]$ such that $(\lambda_i - \lambda_{i-1}) < \varepsilon$, where $\varepsilon > 0$ is an arbitrarily given number. Let $\hat{P}_\lambda$ denote the characteristic function of the compact set $\{M \in \mathfrak{M}_{A_T} : \hat{T}(M) \leq \lambda, \lambda \text{ real}\}$. As in the case of the Riemann integral,

$$\left\| \hat{T} - \sum_{i=1}^{n} \lambda_i^*(\hat{P}_{\lambda_i} - \hat{P}_{\lambda_{i-1}}) \right\|_\infty < \varepsilon,$$

where λ_i^* is any real number between λ_i and λ_{i-1}.

Now since the Gelfand mapping of A_T onto $C(\mathfrak{M}_{A_T})$ is one-to-one, the inverse mapping can be extended from $C(\mathfrak{M}_{A_T})$ onto $\mathscr{B}(\mathfrak{M}_{A_T})$ (where $\mathscr{B}(\mathfrak{M}_{A_T})$ is the class of all Baire functions on $\mathfrak{M}_{A_T}$) because $C(\mathfrak{M}_{A_T})$ is dense in $\mathscr{B}(\mathfrak{M}_{A_T})$ (cf., §33, Chapter VI). Thus, there exist operators P_λ on H such that P_λ is mapped onto $\hat{P}_\lambda$ for each λ. Furthermore, $P_\lambda^2 = P_\lambda$, since $\hat{P}_\lambda^2 = \hat{P}_\lambda$. Since T is the image of $\hat{T}$, under the isometry of $\hat{A}_T$ and $C(\mathfrak{M}_{A_T})$, from the above inequality we obtain,

$$\left\| T - \sum_{i=1}^{n} \lambda_i^*(P_{\lambda_i} - P_{\lambda_{i-1}}) \right\| < \varepsilon.$$

Expressing this inequality in the familiar Riemann-Stieltjes integral form

$$T = \int \lambda \, dP_\lambda,$$

we have derived the main part of the spectral theorem. Finally, it must be remarked that the Bochner-Weil and Plancherel theorem proved in the previous chapter can also be derived from theorems in Banach algebras. The details can be found in the books referred to in the beginning of this chapter.

BIBLIOGRAPHY

1. Abian, A.: The Theory of Sets and Transfinite Arithmetic. W. B. Saunders Company Philadelphia, 1965.
2. Arens, R.: Topologies for homeomorphism groups. Amer. J. Math., *68:* 593–610, 1946.
3. Blumberg, H.: New properties of all real functions. Trans. Amer. Math. Soc., *24:* 113–128, 1922.
4. Bourbaki, N.: Éléments de Mathématique, Topologie Génerale. Livre III. Hermann, Paris, 1940–1949, Chap. I–X.
5. Bourbaki, N.: Espaces Vectoriels Topologiques. Hermann, Paris, 1953, 1955, Chap. I–V.
6. Bourbaki, N.: Théorie des Ensembles. Hermann, Paris, 1960, Chap. I, II.
7. Bradford, J. C., and Goffman, C.: Metric spaces in which Blumberg's Theorem holds. Proc. Amer. Math. Soc., *11:* 667–670, 1960.
8. Cartan, H.: Sur la mesure de Haar. C.R. Acad. Sci., *211:* 759–762, 1940.
9. Chevalley, C.: Theory of Lie Groups. Princeton University Press, Princeton, New Jersey, 1946.
10. Ellis, R.: Locally compact transformation groups. Duke Math. J., *24:* 119–125, 1957.
11. Ellis, R.: A note on the continuity of the inverse. Proc. Amer. Math. Soc., *8:* 372–373, 1957.
12. Fort, M. K., Jr.: Category Theorems. Fund. Math., *42:* 276–288, 1955.
13. Gelbaum, B.: Banach algebras and their applications. Amer. Math. Monthly, *71:* 248–256, 1964.
14. Gleason, A.: Groups without small subgroups. Ann. of Math., *56:* 193–212, 1952.
15. Harr, A.: Der Massbegriff in der Theorie der kontinuierlichen Gruppen. Ann. of Math., *34:* 147–169, 1933.
16. Halmos, P. R.: Measure Theory. D. Van Nostrand Co., New York, 1950.
17. Halmos, P. R.: Introduction to Hilbert Spaces. 2nd ed. Chelsea Publishing Company, New York, 1951.
18. Halmos, P. R.: Finite-dimensional Vector Spaces. 2nd ed. D. Van Nostrand Company, Inc., Princeton, New Jersey, 1958.
19. Hewitt, E., and Ross, K. A.: Abstract Harmonic Analysis. V. I. Springer-Verlag, Berlin, 1963.
20. Husain, T.: Locally convex spaces with the B($\mathscr{T}$)-property. Math. Ann., *146:* 413–422, 1962.
21. Husain, T.: B($\mathscr{T}$)-spaces and the closed graph theorem. Math. Ann., *153:* 293–298, 1964.
22. Husain, T.: The Open Mapping and the Closed Graph Theorems in Topological Vector Spaces. Oxford Mathematical Monographs, Clarendon Press, Oxford, 1965.

23. Husain, T.: Semitopological groups and linear spaces. Math. Ann. *160*: 146–160, 1965.
24. Husain, T.: Almost continuous mappings. Prace Matematyezne (to appear).
25. Husain, T.: Minimal-type locally convex spaces. Math. Ann., *159:* 44–50, 1965.
26. Kakutani, S.: Über die Metrisation der topologischen Gruppen. Proc. Imp. Ac. Tokyo, *12:* 82–84, 1936.
27. Kelley, J. L.: General Topology. D. Van Nostrand Co. Inc., Princeton, New Jersey, 1955.
28. Köthe, G.: Topologische Lineare Räume. 1, Springer Verlag, Berlin, 1960.
29. Loomis, L. H.: An Introduction to Abstract Harmonic Analysis. D. Van Nostrand Co. Inc., Princeton, New Jersey, 1953.
30. Lovitt, W. V.: Linear Integral Equations. Dover Publications, Inc., New York, 1950.
31. MacDuffee, C. C.: Theory of Matrices. J. Springer, Berlin, 1933.
32. Mackey, G. W.: Infinite-dimensional group representations. Bull. Amer. Math. Soc., *69:* 628–686, 1963.
33. Montgomery, D.: Continuity in topological groups. Bull. Amer. Math. Soc., *42:* 879–882, 1936.
34. Montgomery, D., and Zippin, L.: Small subgroups of finite-dimensional subgroups. Ann. of Math., *56:* 213–241, 1952.
35. Montgomery, D., and Zippin, L.: Topological Transformation Groups. Interscience Publishers Inc., New York, 1955.
36. Pettis, B. J.: On continuity and openness of homomorphisms in topological groups. Ann. Math. (2), *51:* 293–308, 1950.
37. Pontrjagin, L.: Topological Groups. Princeton University Press, Princeton, New Jersey, 1946.
38. Pták, V.: Completeness and the open mapping theorem. Bull. Soc. Math. France, *86:* 41–74, 1958.
39. Riesz, F.: Über die linearen Transformationen des komplexen Hilbertschen Raumes. Acta Sci. Math. Szeged, *5:* 23–54, 1930.
40. Robertson, A. P., and Robertson, W.: On the closed graph theorem. Proc. Glasgow Math. Ass., *3:* 9–12, 1956.
41. Rudin, W.: Fourier Analysis on Groups. Interscience Publishers, Inc., New York, 1962.
42. Smithies, F.: Integral Equations. Cambridge tracts in Math. and Mathematical Phys. Cambridge University Press, Cambridge, 1958.
43. Stoll, R. R.: Linear Algebra and Matrix Theory. McGraw-Hill Book Company, New York, 1952.
44. Titchmarsh, E. C.: The Theory of Functions. 2nd ed. Oxford University Press, Oxford, 1939.
45. Van der Waerden, B. L.: Modern Algebra. Vol. I. Frederick Ungar Publishing Co., New York, 1949.
46. Weil, A.: Sur les Espaces a Structure Uniforme et sur la Topologie Génerale. Hermann, Paris, 1938.
47. Weil, A.: L'intégration dans les Groupes Topologiques et ses Applications. Deuxiéme édition. Hermann, Paris, 1953.
48. Wu, Ta-Sun.: Continuity in topological groups. Proc. Amer. Math. Soc., *13:* 452–453, 1962.
49. Hille, E., and Phillips, R. S.: Functional Analysis and Semigroups. Vol. XXXI. Amer. Math. Soc. Colloq. publication, 1957.
50. Rickart, C. E.: General Theory of Banach Algebras. D. Van Nostrand Co., Inc., Princeton, New Jersey, 1960.

INDEX OF SYMBOLS

Set-Theoretic Notations

$\varnothing$	empty or null set
$A \cup B$	union of sets A and B
$A \cap B$	intersection of sets A and B
$a \in A$	a is an element of the set A, or a belongs to A
$a \notin A$	a is not an element of A, or a does not belong to A
$A \subset B$	A is contained in B, or A is a subset of B
$A \supset B$	A contains B
$A = B$	$A \subset B$ and $A \supset B$
$A \not\subset B$	A is not contained in B
$\{x : P(x)\}$	the set of all x satisfying the property $P(x)$
$A \sim B$	$\{x \in A : x \notin B\}$, or the complement of B relative to A
$\{x\}$	singleton or the set consisting of a single element x
(x, y)	ordered pair of elements x and y
$\prod_{\alpha \in A} E_\alpha$	product of sets $E_\alpha(\alpha \in A)$
$f : E \to F$	f is a mapping defined on E with its values in F
$f^{-1} : F \to E$	the inverse mapping of $f : E \to F$, if defined
$f \circ g$	composition of mappings f and g
$1 : 1$	one-to-one
χ_A	characteristic function of the set A
$p_\alpha : \prod_{\alpha \in A} E_\alpha \to E_\alpha$	projection mapping
F^E	the set of all mappings of E into F
$\delta_{ij} = \begin{cases} 1 & \text{if } i = j \\ 0 & \text{if } i \neq j \end{cases}$	Kronecker's delta
$\Rightarrow$	implies
$\Leftrightarrow$	implies and implied by

211

Groups

e	identity of a multiplicative group
0	identity of an additive group
$x + y$	addition of elements x and y of an additive group G
$-x$	inverse of x in an additive group
xy	product of elements x and y of a multiplicative group
x^{-1}	inverse of x in a multiplicative group
$A \pm B$	$\{a \pm b : a \in A, b \in B\}$
AB	$\{ab : a \in A, b \in B\}$
A^{-1}	$\{x^{-1} : x \in A\}$
G/H	the set of all left or right cosets in a group G by a subgroup H
$\varphi : G \to G/H$	the canonical (or natural) mapping

Linear Algebra

$A = (a_{ij}), 1 \leq i, j \leq n$	square matrix of order $n \times n$
$\bar{A} = (\bar{a}_{ij})$	complex conjugate matrix of A
$A' = (a_{ji})$	transpose of matrix A
A^{-1}	inverse of matrix A
$|A|$	determinant of A
$I = A^0$	identity matrix
$\exp(A) =$	$\sum_{n=0}^{\infty} \dfrac{A^n}{n!}$, A square matrix

Topological Spaces

$u = \{U\}$	a topology u, where U is an open set
$u \subset v$	the topology u is coarser than the topology v
$u \supset v$	u is finer than v
$u = v$	$u \supset v$ and $u \subset v$
$\bar{A}$	closure of a set A
$Cl_u A$	closure of a set A in the topology u
A^0	interior of a set A
E_u	topological space E endowed with the topology u
$\mathscr{F} = \{F_\alpha\}$	a filter
$\hat{E}_u$	completion of E_u
$C(E_u, F_v)$ or $C(E, F)$	the set of all continuous mappings of E into F
$C_R(E)$ or $C(E)$	the set of all real or complex-valued, continuous bounded functions on E

$C_\infty(E_u)$	the set of all real or complex-valued continuous functions vanishing at infinity
$C_0(E_u)$	the set of all continuous functions with compact support
$B(E_u)$	the set of all bounded real or complex functions of E_u
$J(f)$	Jacobian of a function f
$\mathfrak{S}$-topology	pp. 33, 168

Topological Groups

x'	character of a group
G'	group of characters or dual group
$\mathbf{M}_n(K)$	the set of all square matrices of order n with coefficients in the field K
$\mathbf{G}_n(K)$	the set of all nonsingular square matrices of order n with coefficients in the field K
$\mathbf{O}_n(K)$	the set of all orthogonal square matrices of order n with coefficients in K
$\mathbf{U}_n$	the group of unitary matrices of order k with complex coefficients
$\mathbf{SG}_n(K)$	special subgroup of $\mathbf{G}_n(K)$ such that the determinant of each matrix is equal to 1
$\mathbf{SO}_n(K)$	subgroup of $\mathbf{O}_n(K)$ having the similar property
$B(\mathscr{A})$ group	p. 90
$B_r(\mathscr{A})$ group	p. 90
$B(\mathscr{C})$ group	p. 89
$B_r(\mathscr{C})$ group	p. 89
$B(\mathscr{K})$ group	p. 97
$B_r(\mathscr{K})$ group	p. 97

INDEX